T0300428

Wireless Independent Living for a Greying Population

Wireless Independent Living for a Greying Population

Lara Srivastava

International Telecommunication Union

River Publishers

Routledge
Taylor & Francis Group

LONDON AND NEW YORK

Published 2009 by River Publishers
River Publishers
Alsbjergvej 10, 9260 Gistrup, Denmark
www.riverpublishers.com

Distributed exclusively by Routledge
4 Park Square, Milton Park, Abingdon, Oxon OX14 4RN
605 Third Avenue, New York, NY 10017, USA

Wireless Independent Living for a Greying Population / by Lara Srivastava.

Routledge is an imprint of the Taylor & Francis Group, an informa business

ISBN 978-87-92329-22-6 (print)

While every effort is made to provide dependable information, the publisher, authors, and editors cannot be held responsible for any errors or omissions.

Dedication

To my parents

Foreword

Evidence of the considerable success of the instinct to survive is widespread. In the case of human society, the growth of medical care, public and private hygiene, and the increase in the availability of relevant knowledge, have all contributed to the phenomenal graying of the population and the lengthening of individual lives in most regions. But as exemplified in the legend of the Sphinx, support is a necessary requirement for the last stage in life (as symbolized by the third leg mentioned in the legend). But it is equally true that while more and more support is required by virtue of an increasingly aged population, less and less of it is available for various reasons.

This book appears at the right moment when several developments have made age and its consequences an important element of human existence. It makes for informative reading, being based on considerations related to independence for the aged and the application of emerging technologies to enhance this independence.

The aim today of most societies, usually expressed through government policies, is to devise methods and institutions to ameliorate the lives of its citizens. The United Nations, through its agencies such as ITU (International Telecommunication Union) and WHO (World Health Organization), has this as an important concern. In one sense or another, this seems to become its very raison d'être. And in pursuit of these objectives, the problems considered and the solutions proposed in this book are timely and relevant.

— Houlin Zhao
Deputy Secretary-General of the
International Telecommunication Union

About the Author

Lara Srivastava is Senior Visiting Fellow at the Center for Tele-Infrastruktur (CTIF) of Aalborg University (Denmark), and policy analyst and program manager of the technology foresight initiatives (previously New Initiatives) of the International Telecommunication Union (ITU). She is responsible for monitoring and analyzing trends in information and communication technology, policy, and market structure. She is the author and manager of publications and research programs undertaken by the ITU, such as the 2005 ITU Internet report on the "Internet of Things" and the 2006 Internet report "digital.life." Prior to joining ITU, she was Consultant and Legal Analyst at Analysys Consulting, Research Fellow in Economics of Infrastructures at Delft University of Technology, and worked in the Legal Department of the Canadian Radio-Television and Telecommunications Commission (CRTC). Lara's main area of interest is technology assessment and foresight — analyzing the impact of new technologies on markets, legal frameworks, and regulation, but also on society and individual user lifestyles. Lara holds a BA Hons and an MA from Queen's University in Canada. She completed an LLB with an emphasis on communications law at the University of Ottawa, studied European law and telecommunications at the University of Paris, and is a qualified barrister and solicitor, member of the Law Society of Upper Canada. She also holds an MSc in science and technology policy from the Science Policy Research Unit (SPRU) of the University of Sussex (United Kingdom).

Acknowledgments

First and foremost, I would like to thank Ramjee Prasad, for his unyielding guidance and encouragement over the last many years. I have learned so very much from him, both personally and professionally, and am grateful for the opportunity he has given me to pursue academic work at Aalborg University.

Thanks to Damir Zrno for his assistance with the computer programming and simulation, his perspicacity and effectiveness. Thanks are equally due to Keith Mainwaring (Cisco Systems) for all the productive brainstorming sessions and technical help. A heartfelt thanks to Katherine Delaney and Yücel Calbay for their loving support throughout.

While at Aalborg University, my research work benefited greatly from discussions with colleagues. Thanks in particular to Horia Cornean, Birthe Dinesen, Frank Fitzek, Flemming Frederiksen, Fred Harris, Persa Kyritsi, Borge Lindberg, Nicola Marchetti, Filippo Meucci, Preben Mogensen, Rasmus Olsen, Petar Popovski, Rocco di Taranto, and Egon Toft.

Preface

grow old along with me!
the best is yet to be,
the last of life,
for which the first was made

— Robert Browning

It is being universally proclaimed that we are at the dawn of a new age —
an age in which information and communication technologies mediate and
pervade almost every aspect of our daily lives. Advances in science are also
enabling us to live longer and healthier lives.

But the extension of life is only half the story. The preservation of the
quality of life is just as important. In general, longevity increases feebleness,
which in turn requires more and more care and support. This support, which
typically needs young muscle, is now in short supply. Something needs to be
done and should be done. Given these changing demographic trends, widening public policy objectives and advances in emerging technologies, the time
seems ripe for devising concrete strategies to improve the living conditions
of an increasingly large population of elderly citizens. This book presents an
independent living home living platform for care, known as the AGE@HOME
platform. The platform is chiefly meant to provide independent living for the
elderly in their own home. But it may be equally used for the physically or
mentally impaired, and applied to institutional care settings.

The AGE@HOME platform aims to provide a home environment that
enables the elderly to live outside institutional care for longer periods. This
latter consideration is seen as increasingly important not only from the individual point of view but also from the societal point of view. The elderly
market has hitherto been underestimated and largely ignored in the process

of technology design and commercial marketing. Social contexts inevitably shape technological diffusion and it is therefore vital for technology designers and service providers to take into consideration important social trends (such as ageing) and their consequences early enough in the innovation process.

Old age is a reality the world over. This has significant consequences on economic growth and societal well-being. It is a fact of life that a greater number of young people will be burdened with responsibilities placed on them both economically and personally by an ageing population.

The platform here presented is a new way of meeting the challenges posed by the requirements of an ageing population, and of giving caregivers more freedom and peace of mind. It goes beyond the rhetoric and introduces a concrete measure toward addressing this important social circumstance.

The wider context of digital living is also examined, with a particular focus on the elderly, covering developments in the open participatory web (e.g. web 2.0) and the expanding network or internet of things. Important demand-side social issues are taken into account, as is the need for promoting from a business standpoint pertinent emerging technologies for the mass market. Socio-ethical considerations are incorporated for a holistic approach to technological design.

This book will be of interest to the general public concerned by problems raised by the ageing of populations. In particular, it will be useful to health professionals, caregivers, information society observers, and experts. The approach taken is multi-disciplinary and multi-faceted, covering not only technical solutions but a wider discussion of the social and human context of technological development in this field.

List of Acronyms and Abbreviations

3G	Third Generation
ADC	Analogue to Digital Converter
CRM	Customer Relationship Management
DARPA	Defense Advanced Research Projects Agency (United States)
DNA	Deoxyribonucleic Acid
DNS	Domain Name System
DSP	Digital Signal Processing
EAS	Electronic Article Surveillance
EHR	Electronic Health Record
EPC	Electronic Product Code
ERP	Enterprise Resource Management
EU	European Union
FDA	Food and Drug Administration (United States)
GDP	Gross Domestic Product
GSM	Global System for Mobile communications
GPS	Global Positioning System
HF	High Frequency
IC	Integrated Circuit
ICT	Information and Communication Technology
ID	Identification
IDT	Interdigital Transducer
ICU	Intensive Care Unit
IEEE	Institute for Electrical and Electronics Engineers
IFF	Identify: Friend or Foe
IM	Instant Messaging
IP	Internet Protocol
IPv4	Internet Protocol Version 4

(*Continued*)

(*Continued*)

IPv6	Internet Protocol Version 6
IPTV	Internet Protocol Television
IR	Infrared
ISO	International Organization for Standardization
IT	Information Technology
ITU	International Telecommunication Union
LF	Low Frequency
MAGNET	My Personal Adaptive Global Net
MIT	Massachusetts Institute of Technology
MMORPG	Massive Multiplayer Online Role-Playing Games
MMS	Multimedia Messaging Service
MSU	Michigan State University
OECD	Organization for Economic Co-operation and Development
ONS	Object Naming Service
OR	Operating Room
PAN	Personal Area Network
PN	Personal Network
PC	Personal Computer
PDA	Personal Digital Assistant
POS	Point of Sale
RADAR	Radio detection and Ranging
RF	Radio Frequency
RFID	Radio-frequency identification
RSS	Really simple syndication
SDL	Specification and Description Language
SMS	Short Message Service
SON	Self-Organizing Network
UHF	Ultra High Frequency
UPC	Universal Product Code
UWB	Ultra Wide Band
VHF	Very High Frequency
Wi-Fi	Wireless Fidelity
WiMAX	Worldwide Interoperability for Microwave Access
WLAN	Wireless Local Area Network
WMAN	Wireless Metropolitan Area Network
WPAN	Wireless Personal Area Network
YRP	Yokosuka Research Park

Contents

1

Introduction

The older the fiddler, the sweeter the tune.

— English proverb

1.1 Life in the Years

"In the end, it's not the years in your life that count. It's the life in your years." So said Abraham Lincoln and he was not far off the mark. Although many things in life are uncertain, the fact that we grow old is a stark certainty. And today, we are growing even older, enjoying longer lives afforded to us by better health care and quality of life. At the same time, the number of young people to look after us as we grow old is declining, and entire populations are ageing across the globe. So the question is — how do we ensure that life truly remains in the years?

There is a growing need to address the living conditions of the aged in many industrialized societies. Current health care and pension systems are already heavily burdened, and this situation will only worsen as the number of elderly increase over the next decades. At the same time, single member households are on the rise: one out of eight persons in the European Union is a single adult living alone [1]. It is in the interest of society to provide its elderly with independence and self-sufficiency, not only as a cost-cutting measure, but also for motivating and encouraging individuals to remain productive members of society. If today, as they say, the 60s are the new 50s, tomorrow they may be the new 40s.

This book is primarily concerned with the amelioration of living conditions for an increasingly aged population through the development of a technical platform to assist the elderly in their own homes. It is motivated by the growing need for multi-disciplinary approaches to technology design. It also takes into account that technological innovation and development should not

1

only occur for the sake of advancing the state of knowledge as a whole, but should also seize opportunities to address the needs of particular users — their particular set of requirements and preferences. This holds even more true for information and communication technology (ICT), because no other technology comes closer to the very core of human existence: it now mediates many forms of social interaction, has become key to business interests and is vital for public health and safety worldwide. Moreover, with the advent of the internet and particularly the mobile phone, ICT has become the most intimate form of technology that we use today. Given its vital role in daily life and society, together with its potential for health and safety applications, ICTs present an extraordinary opportunity to enhance the overall quality of life for the aged.

1.2 Challenges

In the context of ageing and the use of ICTs, there are a number of challenges to be faced. Demographic trends are straining public health care systems, as more and more people require sheltered or nursing home care and can no longer feel independent enough to live in their own home. At the same time, technologies for making the home more network-enabled, i.e., efforts toward creating the digital home, are suffering from slow take-up and lack of economies of scale. Technologies like radio-frequency identification (RFID), which hold great promise for data transmission, monitoring, and sensor integration, are not being adopted at a rapid rate, due to, *inter alia*, the fragmentation of applications in the market, the lack of "killer" user-driven applications, the lack of awareness and even reticence among users. The AGE@HOME platform presented here seeks to address these challenges, in the following manner:

- The platform is intended to enable the elderly to live independently longer, in their own home, outside of costly nursing home care facilities, through the use of nonintrusive wireless technologies.
- The platform monitors behavior and environmental conditions in the home, to alert of any crisis situations, thereby making the elderly feel safer when living on their own.
- The platform provides a simple solution for equipping the home with promising emerging technologies, and as a result, offers an

early unobtrusive digital home environment that can be built upon in future projects.

- The deployment of the platform on the market has the potential to create specific market demand that would lessen user concerns surrounding so-called "monitoring" or "surveillance" technologies like RFID.

There are a number of other related challenges that need to be addressed but lie outside the scope of this study. First, home health data and records must be better integrated with hospital and clinic electronic health records. Standardization of RFID technology in and out of the digital home space requires further global attention. Further work is required on the development of biomedical sensors for home care and their integration with hospital platforms. Indoor localization techniques based on wireless technology require further sophistication. Security issues related to RFID and sensors must be dealt with, particularly in a global environment that poses increasing challenges for data security and privacy. Finally, overall system affordability (for individual and institutional users) must be promoted in order to stimulate mass adoption. User acceptance is vital for any system to succeed on the market.

1.3 About the Book

In the main, this book investigates the possibility of adapting emerging wireless technologies to the home for monitoring the behavior and safety of elderly residents. Within the context of the ageing phenomenon and independent home living, it is concerned with research problems that relate to network design, technology, and user requirements.

The AGE@HOME platform introduced herein is meant to enable elderly people, particularly those residing alone, to live independently in their own home through the assistance of wireless technology. It is designed to alert caregivers of potential crises in the home, in order to enable safe, secure, and autonomous living. The main target group of AGE@HOME is the aged. However, the platform would be equally suited to persons with physical or mental disabilities.

The AGE@HOME platform attempts to combine and integrate various emerging technologies with more traditional technology in order to create what may be seen as an early prototype of a digital home for the aged. Such a digital environment must be achieved without restricting resident activity through the over-monitoring of daily life, undue interference or the use of intrusive or bulky devices.

This book relies on extensive research in the fields of innovation management, technology policy, and wireless communications conducted over the last five years. The research is based on a survey of recent public domain sources, academic journals and publications across various disciplines, ranging from engineering and telemedicine, to innovation management and sociology. Informal interviews with industry players and elderly individuals were also conducted.

Following the research phase, a technical home platform to address the challenges and requirements of the elderly was conceptualized. Specification and Description Language was used to conceive the decision logic of the system prior to the programming phase. Based on that logic, a software program was developed, and a random simulation model was generated to test the software.

The author presents these specific technical aspects of the platform while exploring the larger context of ongoing technological evolutions, e.g., web 2.0. It includes a much-needed holistic approach to the human being as a user of technology, in line with the view that technological design should be based on a better understanding of societal contexts and incorporate multi-disciplinary perspectives. For this reason, the book begins with an examination of social trends and structures in Chapter 2. Chapter 3 proposes a new theoretical framework for use leadership in technology design and innovation before launching into the specifics of the technologies under consideration in Chapter 4. Chapter 5 presents an overview of the AGE@HOME platform for independent living and its main objectives. Chapter 6 provides a graphical description of the platform, describes the software implementation, and presents the results of the random computer simulation. It also summarizes the challenges that are raised. Chapter 7 explores the wider context of digital living, including socio-ethical implications. Conclusions and future directions are presented in Chapter 8.

2

Emerging Technologies and the Aged Market: Innovation in Its Social Context

With mirth and laughter let old wrinkles come.

— William Shakespeare

Technology does not exist in a vacuum [1]: social contexts constantly shape technological diffusion. Conversely, technological innovation and adoption affect society. It is therefore vital for technology designers, service providers and policy-makers to take into account existing and relevant social trends and the effect of new technological developments on them, when innovating, drawing up business plans, and promulgating regulation. The graying of the population is one such important trend. This chapter examines current technological developments in this and other social contexts.

2.1 Beyond the Digital

In the domain of technology, we are face-to-face with a full-blown digital revolution. Everyone and everything is going digital: most of us spend hours a day in front of a screen, sending and receiving bits and bytes of data. At home, we watch programming, send emails and photos, listen to music, bank or make travel plans, all through the use of digital technology. Digital media are rapidly replacing all other forms.

It would be insufficient, however, to say that we are merely converting our analog world into a digital one. It is a much more complex and multi-faceted journey. There are a great number of technological developments and trends that are radically changing our relationship to technology, to our physical surroundings, and to each other:

- technical, corporate, and regulatory convergence
- service mobility and the growth of wireless networks

- growing value of real-time, "always-on" information & communication
- popularity of portable ICT (information and communication technology) devices
- advances in computing rendering information "ubiquitous"
- advances in miniaturization
- growth of social networking & web 2.0 apps
- continuing expansion of the internet as a platform for information, communication, collaboration, entertainment, and creativity
- growth of user-generated content and value

But perhaps the most overarching trend is that we are no longer able to separate technology from daily life. It has become part and parcel of it. Moreover, the speed at which innovations are making it to market means that often, their impact on society is a *fait accompli* before it is weighed. This is why it is more important than ever before for technologists, designers, businesses, and regulators to gauge the impact of technology on society, and incorporate their analysis into public policy and commercial strategy.

2.2 Convergence at the Core

The phenomenon of convergence, though not entirely new, has become the subject of growing interest worldwide. It is viewed as an underlying trend in the information and communication industry that is fuelling innovation and diversifying services. The European Union (EU) Seventh Framework Programme defines technological convergence as follows: "convergence (converged environments/networks) defines a multimedia environment and/or network where signals regardless of type (i.e. voice, quality audio, video, data, etc.) and encoding methodology may be seamlessly exchanged between independent endpoints with similar characteristics [2]." In other words, convergence implies that previously vertically integrated services, such as multimedia, voice and data services, are being provided on common platforms (e.g. internet of mobile networks). For example, due to developments in component miniaturization and low-cost processors, it has become possible for mobile phones to support a full range of voice, video, and data services.

The term convergence applies at many levels. It not only applies to technology (e.g. network and device convergence), but also to regulation, content/services, and corporate structure. In response to technological convergence, regulatory structures are trying to adapt, with more and more governments aiming to merge their broadcasting and telecommunications regulatory frameworks. Industrial convergence is evident, for instance as large media companies merge with telecommunications enterprises, and computing is increasingly combined with consumer electronics.

As it is more and more difficult to distinguish between voice, data, and broadcast networks, differences between players are becoming more apparent. Fixed operators are typically bundling their services in an effort to expand their portfolios and retain customers. Mobile operators are promoting mobility as a key network element and marketing it as a replacement or complement to fixed services. Regulators are struggling with the question of how to regulate similar services provided over different platforms (e.g. broadcast television versus television programming provided over the internet). Internet players like Google have entered the business of voice communications with Voice over Internet Protocol and telecommunications providers, like BT, are keen on making profits from the entertainment business with Internet Protocol Television. In the meantime, for consumers, the search continues for simple, affordable, and seamless services regardless of the device they use to access them.

Although convergence has given rise to a number of new technical and regulatory challenges, it has also provided the possibility of making communications cheaper, more accessible and portable. It has also stimulated innovation and created a multitude of new service opportunities. Globally, convergence will come in many flavors and the degree of convergence will vary from country to country. Consequently, different business models will have to co-exist for some time, and regulation must be in place to ensure that open market structures foster innovation and industry collaboration.

2.3 Social Trends and Tensions

"Change is the only constant," says the Greek philosopher Heraclitus. Never before have changes tumbled in such rapid succession as in the present age. These changes have not only occurred in industrial and commercial arenas, but also, just as importantly, in social and societal structures.

Let us examine some of the more noticeable social trends today. It may be useful to consider these trends in the form of tensions or transitions. In some cases, there is a stark transition from one way of doing things to another, or from one viewpoint to another. In others, there is a tension or conflict between two different ways of doing things or between two different perspectives.

1. *Relative predictability–unpredictability (transition)*
   Over the past decades, our environment has transitioned from one of relative stability to one that is increasingly unpredictable. Market environments, in particular, are difficult to predict as are user demand and preference. The wealth of choice individuals have today means lifestyles, career development, and social positioning are evolutionary rather than static. In the past, one's place of birth mostly determined one's place of death and mode of life. Today, this is not so.

2. *Less choice–more choice (transition)*
   Most individuals have much more choice than they have had in previous generations: choice of where to live, what/when and how to study, work or travel, whom to communicate with, when to communicate and how, and how and what to buy. This can empower individuals and provides many social advantages. But choice also represents a burden, with some finding it is increasingly difficult to make decisions, commit to action, or understand and accept consequences.

3. *Information-poor–information-rich (transition)*
   A plethora of information is now available at our fingertips, anytime and anywhere. It is as simple as a click of a mouse to access information about events occurring across the globe, instantaneously. It is also much easier to find information about topics of interest, without having to travel to a library or consult copies of newspaper archives.

4. *Time-rich–time-poor (transition)*
   A wealth of information and penury of time is what characterizes our age. Although technology is intended to provide convenient tools to help us save time, it seems that overall leisure time has decreased. The rapid pace and pressures of modern life are

compounded by the unpredictability of relationships, both professional and personal, and an uncertain global economy. There is a continuing dissatisfaction with human relationships. Finding the time to spend with the family, and the disposition, is hard to come by, notably in Europe and North America.

5. *Full-service–self-service (transition)*

In many areas, the convenience and control of the automated transaction (e.g. e-banking) has made life easier. Businesses, too, favor the lower cost of operations and the greater geographical coverage. However, this trend is not limited to technology applications. In general, the "do-it-yourself" trend has gained momentum in many aspects of daily life, from public utilities to retail.

6. *Fixed–mobile (transition)*

Due to changes in technology, there is an undeniable shift from fixed to mobile services. More and more networks integrate mobility as a central element in design, and a larger number of users prefer mobile devices. Individuals are also more socially and geographically mobile: they travel and work in different countries and evidence more social and career mobility.

7. *Passive user–active creator (transition)*

Services are becoming more personalized and individual preference is playing an important role in the way that individuals communicate, learn, and access information. With the more open and participatory approaches to the World Wide Web, internet users are no longer limited to consulting information but have become instrumental in generating content themselves, thereby enhancing the overall value of the networks they use.

8. *Real–virtual (tension)*

As people spend more and more time online, and as data about our physical world becomes readily available in the virtual world, real and virtual contexts will complement each other, but may also collide with each other.

9. *Public–private (tension)*

With the widespread availability of information and communications, the boundaries between the public and private spheres of human existence are beginning to blur. Private mobile conversations occur in public spaces. Similarly, all kinds of private data are now stored (and even posted) on public networks, e.g., through email, tracking of surfing habits, and social networking sites such as Facebook [3]. What used to be private has become public. Thus, the public domain now contains a greater and growing amount of private data.

10. *Collective–individual (tension)*

    On the one hand, modern society favors individualism. On the other hand, the internet and tools like online social networking have enabled an entirely new form of sharing and collaboration. There is tension between the need to retain individuality and individual self-expression and the requirement to conform to a group, a point of view, or a common (digital) culture.

11. *Young–old (transition)*

    In general terms, the world's population is ageing, with Europe being no exception. Over the next few decades, an increase in the average age will mean that a greater number of people will require assistance in daily life, and greater burdens will be placed on health care and pension systems.

Understanding the above tensions and transitions in daily life can serve to inform and enhance the development, adoption, and use of new technologies for the public interest.

2.4 Strategies for the Public Sector

Many public sector institutions are looking at information and communication technologies as a way to improve the conveniences of daily life. They are exploring ways to streamline government services through greater automation and on-line information and processing, thereby aiming to further empower the citizen. Governments are also concerned that the adoption of new technologies is too rapid for legal and regulatory frameworks to adapt, and can render them ineffective. At the same time, the overzealous use of regulation might hinder the creation of new markets and stifle innovation. The protection of personal privacy and the existence of cybersecurity threats are of particular concern. There is a need, therefore, for governments and regulators to work with individual players and users in devising the network of tomorrow — a network consisting not only of people, but devices and things (i.e. the network of things). This future network should be global, interoperable, scalable, and sustainable.

In parallel with these objectives, governments aim to put into effect social policies that lead to greater employment, better education, heightened public safety, and excellence in health care. There now exists a clear advantage in applying emerging technologies in these domains. Activities suffering from a lack of resources, such as education and health care, are particularly suited and in this context, private–public partnerships are a viable option to promote innovation. This approach will most likely spread the benefits of broadband access to a wider proportion of the population, which in turn, is likely to have a bountiful effect on growth and productivity.

2.5 A Question of Demographics

Demographic trends over the last several decades clearly indicate an ageing of the world's population. While medical innovations have increased life expectancy and enhanced the quality of life, fertility rates have declined. In Europe for instance, the average age is set to increase from just under 39 in 2000 to around 48 in 2050 (Figure 2.1). At current rates, the old-age dependency ratio (i.e. ratio of the population aged 65 and over to the population aged 20–65) will reach from 35 percent to over 50 percent by 2030, and from 40 percent to over 70 percent by 2050 [4]. Although there is some variation in the rates of ageing around the world (i.e. ageing is more acute in Japan and in European countries like Italy and Germany than it is in Canada or the United States), it remains a universal phenomenon.

Fewer young people translates into reduced resources and slower economic growth. In some industrialized countries, the labor force is set to decline [6]. Between 2000 and 2050, the ratio of inactive population aged 65 and over to the workforce is expected to increase dramatically in OECD (Organization for Economic Co-operation and Development) countries: in Korea, for example, the ratio is predicted to increase sevenfold. This is in sharp contrast to some emerging economies, such as India [7] and Brazil, where young people are not in short supply (Figure 2.2). In fact, world population growth is taking place primarily in developing countries [8]. In the industrialized world, however,

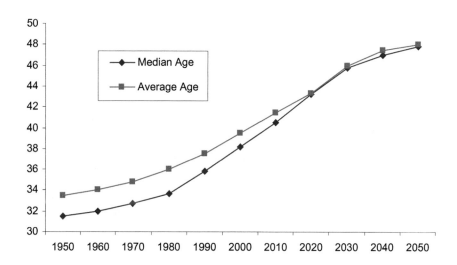

Fig. 2.1 Median and average age in EU 27 (1950–2050).
Source: European Commission [5].

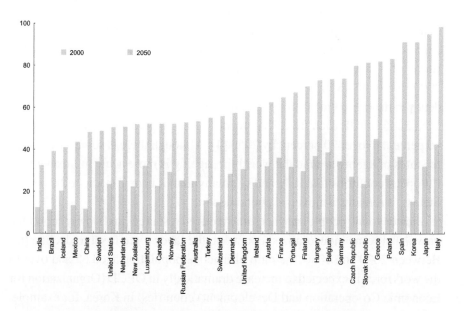

Fig. 2.2 Ratio of inactive population aged 65 and over to the workforce.
Source: OECD Factbook 2008 [10].

population ageing has resulted in significant fiscal consequences on health and pension systems, a situation which is likely to worsen in the future. More young people will be required to take on caregiving roles while a greater number of aged persons will become dependent on caregivers and public health systems. A key fact to be observed is that more and more elderly people are living alone: in the European Union, for instance, 30 percent of women and 13 percent of men between the ages of 65 and 74 live alone [9].

As a result, there is an urgent need, notably in Europe, to develop innovative and independent platforms for the care of the elderly. It is in society's interest, as a whole, to facilitate lifelong learning and to create an environment that promotes a constructive, enjoyable, safe, and secure "golden age." Information and communication technologies (e.g. mobile wireless communications, internet) already pervade many aspects of our daily lives, and can be important enablers for ensuring quality of life for the aged. More specifically, innovations in the area of short-range radio technologies and wireless sensor networks may be vital to the development of real-time public safety and welfare applications.

2.6 Seizing a New Opportunity

Against the backdrop of the social trends described above, it is important to pay due attention to a market that has hitherto been underestimated and largely ignored in technology design and commercial industry marketing: the aged.

Targeting the aged as an important market segment is not only vital for society and social welfare, but it also represents important business opportunities for equipment vendors, service providers, and innovators. Quite beyond demand-pull strategies, technology for the aged is a lacuna in technological development that is waiting to be filled.

Noninstitutional caregivers are income earners and have a vested interest in ensuring the care of their ageing loved ones. They may be directly or indirectly influenced by the consumer habits of the people they care for. If, for instance, they see a certain gaming application which is appreciated by a family member, they may be inclined to buy it for themselves, either now or in the future. Both the elderly and caregivers may act as "opinion leaders" in this respect. One might draw inspiration from the "lead user" [13] theory in innovation: this group stands to benefit substantially from obtaining a solution to their needs, and with respect to creating networked digital homes, they can act as an important needs-forecasting mechanism [14].

Encouraging penetration of certain technologies among the aged could serve to dispel some of the myths about surveillance technologies and counter some of the negative press that has plagued technologies such as RFID. Media hype could be replaced by practical solutions that would be seen as publicly beneficial. The use of these technologies in the home could create a paradigm shift in mass usage, as it did with the internet: independent living platforms could be prototypes for the digital home of the future.

This approach fits squarely in line with the social trends and transitions mentioned above, in particular:

- people have less time;
- people are more mobile — they travel and change residences more often;
- there is a trend toward greater individualism and independence;
- society is increasingly applauding self-sufficiency and self-service.

In this context, one wonders who will look after us as we age. Despite an unpredictable and unforeseeable future, there is one thing that is undeniable: all of us, as individuals, will age.

If one considers today's public policies on emerging technologies in light of current social trends and long-standing social policy (e.g. health care), it would seem that the time is now ripe to emphasize applied research on the use of emerging technologies to address the ageing phenomenon.

This book attempts to seize this important opportunity. It puts forward a technical platform specifically designed to meet the requirements of the elderly at home. Many challenges lie ahead in the field of health care, public welfare, and the care of the aged. Social policies will likely need to be re-aligned as a result. The AGE@HOME platform proposed here is acutely relevant to these challenges and policy issues.

3

Toward a New Theoretical Framework for "Use Leaders"

Once you're over the hill, you begin to pick up speed
— Charles M. Schultz

As mentioned above, traditionally stable market environments have been giving way to environments where unpredictability and change have become the norm. As a result, industry faces increasing competitive pressures not only from other firms but also from rapid and often unforeseeable technological change. These pressures are only exacerbated by the global nature of markets and the knowledge-based economy. Firms must compete in their capacity to innovate and to retain "knowledge-intensive" skill sets, but also in their ability to understand increasingly heterogeneous user demand and requirements. Such an understanding of demand must go beyond mere market research for it to be most commercially effective. There is recognition among industry players that a greater emphasis on so-called "demand-pull" considerations is required. By being unduly focused on technology alone, firms risk becoming out of touch with markets, developing products that are little used, while neglecting to develop or leverage products with mass market potential (technology-push). At the same time, of course, being solely driven by market demand runs the risk of stifling innovation (demand-pull). As Henry Ford put it, "if I had asked end customers, all they would have said is that they need a faster horse." Ideally, a balance must be struck between the technology-push and the demand-pull strategies. This is the rationale today behind user-centric innovation and its subset, user-driven innovation.

3.1 Users Take Center Stage

Steve Jobs, the founder of Apple, once said "being innovative in an organized and user-friendly fashion is the essential comparative parameter of the

21st century." User-driven innovation is a systematic approach to developing new products and services, based on user requirements, practices and identities. The term "user-driven innovation" has been increasingly used to denote a shift in focus to demand-pull strategies rather than emphasizing technological development alone. The methods for user-driven innovation are diverse, ranging from simple surveys and interviews, to the creation of prototype products for testing by user groups (e.g. focus groups). Despite the increased attention on user demand, the term "user-driven innovation" remains relatively indistinct and is often used interchangeably with user-centric innovation. These two terms, however, must be distinguished. User-centric innovation is based on the needs of users in a wider sense — it is an umbrella term under which more specific approaches to user-driven innovation can be developed. All user-driven innovation is user-centric, but not all user-centric innovation is necessarily user-driven.

In the ICT industry, fast-paced technological developments together with unpredictable markets means that it is becoming more and more difficult to suitably evaluate demand for particular new products and services. As a result, firms have begun to place a greater emphasis on demand assessment. Business development and corporate strategy departments are working hand in hand with marketing departments to ensure that the needs of users are taken into account as a business priority. But what has now emerged as popular industry jargon (e.g. "user-driven," "user-powered") does not necessarily strike the much-needed balance between demand-oriented and technology-oriented business and product development. Mere market research and user surveys may not generate enough knowledge to inform technology designers and strategists in unpredictable market environments, particularly in the light of increasingly heterogeneous nature of demand in a globalized economy.

User-centric innovation has a number of important benefits, namely:

- it fosters "learning organizations";
- it helps maintain a firm's competitive lead, provides advance warning of evolving needs of the broader market;
- it channels user needs and improvements back into products;
- it stimulates market adoption (avoiding scenarios in which products or services are looking for a market);
- it encourages open & collaborative business environments; and
- it introduces a diversity of perspectives in design.

This attention on the user during the innovation process has also been called "democratizing innovation [1]," as it promotes multi-disciplinary thinking and open innovation processes, rather than closed ones that prioritize the protection of investments. Finally, and perhaps most importantly, an emphasis on user interests means that the human being becomes part and parcel of the innovation process. This can only stimulate individual well-being and social welfare, benefiting overall societal development and economic growth.

3.2 Lead Users and the Innovation Process

Firms are concerned with the adoption of their products and services in the marketplace, and with their rate of diffusion. The theory of the diffusion of innovations was first introduced by Everett Rogers in 1962 and he defined it as "the process by which an innovation is communicated through certain channels over time among the members of the social system."

Much research and analysis has since been conducted on the diffusion of innovations. It is now well-established that new technologies or products typically diffuse through a society over many years, rather than having a simultaneous impact on all members of a society. There are early adopters, that tend to adopt technology earlier than the majority, and but also laggards, who tend to be late adopters of technology. Everett Rogers popularized the famous bell curve shown in Figure 3.1.

Different socio-economic groups have conflicting interests and motivations, in terms of their likelihood to adopt technology. Different choices and decisions regarding technology affect them differently [3]. According to Rogers, the attitude of people toward new technologies is a key element in its diffusion. In this context, he identified those characteristics of innovations, as perceived by adopters, which would increase the likelihood that they will adopt a technology:

- *relative advantage* (the degree to which an innovation is perceived as being better than the idea it supersedes);
- *compatibility* (the degree to which an innovation is perceived to be consistent with the existing values, past experiences, and needs of potential adopters);
- *complexity* (the degree to which an innovation is perceived as difficult to use);

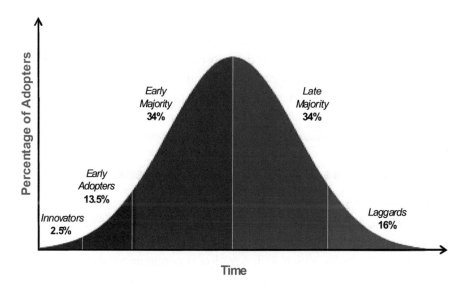

Fig. 3.1 Diffusion of innovations and categories of adoption.

Source: Adapted from E. M. Rogers [2].

- *trialability* (the opportunity to experiment with the innovation on a limited basis); and
- *observability* (the degree to which the results of an innovation are visible to others).

In order to better understand what types of users might prompt marketable and innovative products, Eric von Hippel first expounded the lead user theory in a 1986 paper entitled "Lead users: an important source of novel product concepts [4]." In this paper, von Hippel theorizes that identifying "lead users" of a product or process can be very useful in informing the design and development of new product concepts. He states that current market research is typically insufficient in the case of product categories characterized by rapid change, for instance high-technology products as explored in this book. Von Hippel proposes a solution with the identification of a particular set of users defined as those whose present and strong needs foreshadow the needs of the general marketplace. These "lead users" are typically familiar with conditions that lie in the future for most other users. In contrast to the ICT or high-tech industries, most consumer products are relatively slow-moving, e.g. dishwashers or varieties of bread do not differ radically from version to version. Therefore, understanding the requirements of the typical user may suffice. However, in rapidly changing environments, such as the ICT industry, the day-to-day experience of the average user may not always remain valid during its lifecycle or even by the time a product is developed. Lead users may offer particular insight and serve to stimulate the eventual adoption of an innovation. Von Hippel defines lead users as having two main characteristics:

(a) *"lead users face needs that will be general in a marketplace but face them months or years before the bulk of the marketplace encounters them";*
(b) *"lead users are positioned to benefit significantly by obtaining a solution to those needs."*

Lead users refer mainly (though not exclusively) to firms in this context, rather than to end users. However, this approach can also be effectively used to identify lead end users in the case of particular innovations. It should be noted that lead users as defined by Eric Von Hippel are not to be confused with early adopters of an innovation, as defined by Rogers. Lead users can in fact be innovators in this context and can be seen to precede early adoption: they are user-innovators or user-producers of products or processes (Figure 3.2). The main thrust of the argument is as follows: first, markets tend to evolve along certain underlying trends; second, lead users experience needs earlier than the rest of the market; and finally, their response to those needs can improve the attractiveness of particular innovations [5].

Von Hippel suggests that the notion of lead users can be used in marketing research, using a four-step process. The first step is to identify an important trend in the market. The next step is to identify lead users who are able to obtain a relatively high net benefit from adopting a solution to needs that are related to those trends. The third and fourth steps are to analyze the lead user data, and project that data onto the general market of interest.

Although von Hippel's theory has been understood mainly as relating to leading "firms" in a particular market, it provides, in conjunction with Rogers' theory of innovation, an excellent basis for developing a framework for the leadership of end users in the diffusion of innovations.

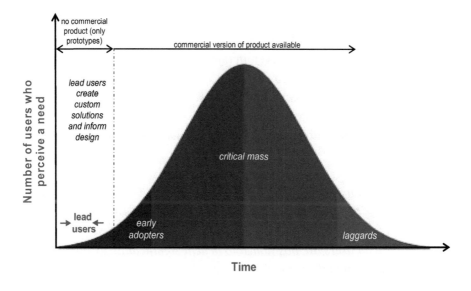

Fig. 3.2 Lead users and innovations.

Source: Adapted from E. von Hippel [6].

3.3   Chasing Demand: The Use Leader Framework

Whereas lead users are seen to be ahead of the entire adoption curve as innovators, "use leaders" are end users that fit into the category of early adopters that have an influence on the overall market. In the case of digital homes, early adopters are typically hobbyists with large disposable incomes. But what is of interest to the rate of diffusion is the role of use leaders with a significant influence on the market. Although they may not innovate themselves (like von Hippel's lead users), they can serve to inform design and influence market adoption as a whole. Once they adopt a technology, they might serve as opening leaders that could serve to accelerate the diffusion process of new products [7].

An approach based on "use leadership" incorporates the characteristics of users as well as the characteristics of the technology or service, i.e. both demand-side and supply-side considerations. It is a flexible approach, in that one may identify either innovations or users in the first instance.

Use leaders are typically an identifiable social grouping. These groupings can be easily identified on a factual basis, and do not include groupings based merely on variable taste and opinion. They represent a discrete group of people that do not change significantly over time. Examples include wheelchair users, the blind, the elderly, teenagers, jobseekers or homeless. This categorization is not dissimilar in nature to the "trait cleavages" described by Douglas W. Rae and Michael Taylor in "The Analysis of Political Cleavages [8]." In political science, the term "cleavage" is often used to describe the factors determining the behavior of voters in elections. Rae and Taylor defined three categories of "cleavages": "trait cleavages" such as gender, religion, ethnicity, and class; "attitudinal cleavages" or differences based on opinion; and "behavioral cleavages" such as membership of organizations. Although Rae and Taylor's classification has been subject to criticism when applied to voter allegiances, the empirical characteristics of cleavages are a useful way of defining the nature of "use leaders" and can better inform the identification of relevant social groupings. Once a grouping is identified (e.g. the elderly as a group), certain criteria must be fulfilled before that group can qualify as virtual "use leaders" of a new technology, product or service.

In particular, use leaders must possess the following characteristics (i.e. demand-side characteristics), the first two of which are similar to von Hippel's definition of lead users (Figure 3.3):

1. *They have **specific needs** and stand to benefit substantially for finding a solution to those needs*;
2. *They experience **early needs** that the market-at-large has yet to experience and may never experience until solutions become available (e.g. the question posed by Henry Ford as mentioned above)*;
3. *They have a direct of indirect **influence on the spending decisions** of individuals or organizations with disposable income.*

Second, on the supply side, the characteristics of the innovation itself are as follows:

4. *The innovation **empower**s users — finding a solution to their needs will empower them in their daily lives*;

and/or

5. *The innovation serves a **greater social purpose** beyond the individual, i.e. a greater socio-political objective is achieved.*

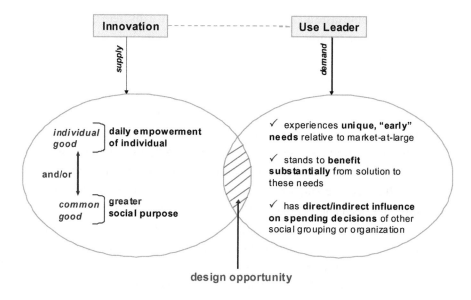

Fig. 3.3 Use leaders and innovation.

The success of an innovation will be even greater if both the individual and common good are combined: user empowerment and achievement of a socio-political end. Finding a solution to the needs of "use leaders" will typically have a spill-over effect to other social groupings and informal networks. Use leaders, by the very nature of their needs and their role in society, can exert in influence the marketplace. This contagion process is enabled by their source as communication sources [9].

The case of the elderly and digital home technologies provides an excellent example of user leadership and this is discussed in the section that follows.

3.4   Applying Use Leadership

In the present book, it is argued that members of the aged community fulfill the criteria to be use leaders in the field of digital home technology. First, they experience unique needs: physical frailty combined with a strong desire to be independent in their own home, while feeling secure that any crisis situation can be mitigated. Second, they would benefit substantially from digital home technology that would address these requirements of independence and safety. Third, the aged have a significant influence on the spending decisions of their caregivers, particularly members of the same family or community, who have a vested interest in ensuring the well-being of their loved ones, while easing their care-giving responsibilities.

Moreover, those aged 65 and above tend to have high disposable incomes themselves and therefore are an important target market for all players in the consumer goods area, including consumer electronics. In Japan, for instance, pensioners in 2005 possessed almost half of the country's 11.3 trillion yen in savings, and had an income only 10 percent less than the national average [10].

In this particular case, the innovation in question, digital home monitoring, also fulfills the criteria of use leadership on the supply side. It empowers the individual in their day-to-day activities (individual good) and also serves a greater social purpose (common good): it enables the elderly to live outside institutional care, and remain independent for longer, while addressing societal concerns related to an ageing population.

3.5 In Pursuit of the Individual and Common Good

As mentioned above, the care of the elderly through the use of emerging technologies will not only benefit the individual good, but will equally contribute to a common good in society. The responsibility that society has vis-à-vis its vulnerable members will be better fulfilled, in a way that is respectful of human dignity. It has been said that the more civilized a society is, the better its inclusion of marginalized groups.

Using technology to enhance the quality of life of the aged is both a business and social opportunity. In this sense, both the common good and collective good are served. Though an important objective, collective good is mainly material — it is divisible: the more one individual has, the less the other has access to [11]. For example, if a society's financial resources are directed to assist a particular group of people, this leaves fewer resources for other groups. Common good, on the other hand, is diffusive or self-replicating: its very existence encourages and facilitates an increase in itself. For example, peace in a society is a common good. Privacy is an important common good. Similarly, care and respect for the elderly in a society instils desired values in, and promotes quality of life for, other members of the population. It assures them that they will receive the same respect as they grow old, thereby alleviating anxiety and depression in the overall population, while fostering independent thought and self-expression. Using technology to enhance the quality of life of the aged is both a business and social opportunity.

3.6   In the Marketplace: From Hype to Usage

Encouraging penetration of certain technologies among the aged, as a key "use leader" group, could serve to dispel some of the myths about surveillance technologies and counter some of the negative press that has plagued technologies such as RFID. Media hype (see also Section 4.1.4) could be replaced by practical solutions that would be seen as publicly beneficial. The use of these technologies in the home could create a paradigm shift in mass usage, as it did with the internet: independent living platforms could be prototypes for the digital home of the future.

This approach fits squarely in line with the social trends and transitions mentioned in the previous chapter, in particular:

- people have less time;
- people are more mobile — they travel and change residences more often;
- there is a trend toward greater individualism and independence;
- society is increasingly applauding self-sufficiency and self-service.

In this context, one wonders who will look after us as we age. Despite an unpredictable and unforeseeable future, there is one thing that is undeniable: all of us, as individuals, will age.

If one considers today's public policies on emerging technologies in light of current social trends and long-standing social policy (e.g. health care), it would seem that the time is ripe to emphasize applied research on the use of emerging technologies to address the ageing phenomenon.

The present book seizes this important opportunity. It puts forward a technical platform specifically designed to meet the requirements of the elderly at home. Many challenges lie ahead in the field of health care, public welfare, and the care of the aged. Social policies will likely need to be re-aligned as a result. The AGE@HOME platform proposed herein is most relevant to these challenges and policy issues.

4

RFID in Focus: Technology, Use, and Complementarity

Life can only be understood backwards, but it must be lived forwards.

— Søren Kierkegaard

RFID is a wireless technology suitable for low-power transmission of data within the home. It is also vital to the expansion of tomorrow's networks, which will connect not only people and computers, but also objects and things in the surrounding environment. The chief strength of RFID is that it is not only a short-range wireless transmission technology but also has the capability to identify unique items, and as such, it holds great potential for managing the identity and location of people, devices, and things.

In addition, RFID is increasingly being used in health care applications, and therefore, integrating it within the home can serve to facilitate the much-needed convergence between home care and hospital care environments. RFID is one of the key components for the AGE@HOME platform as described in Chapters 5 and 6.

This chapter discusses the deployment of this important emerging technology, including market developments and technical issues. It explores the origins of the technology, its applications, advantages, and inherent limitations. It also examines its deployment in health care and summarizes important gaps before examining its combination with sensor technologies and exploring future prospects.

4.1 RFID Development and Its Technical Aspects

4.1.1 Origins and Development

The concept of radio-frequency identification is not new. Radio itself, of course, dates back to early 19th century work on electromagnetic energy by Michael Faraday, and the discovery of radio waves by Heinrich Hertz in 1887. In the early 20th century, efforts began to develop radio systems that would enable the transmission and reception of radio waves: radio detection and ranging (RADAR). In World War II, many countries used RADAR to navigate ships and detect enemy craft, but this was a simple noncooperating use of radio waves, i.e., ships, for instance, could be detected without their active participation in the process. Around the same time, work began on the practical use of radio waves to detect and locate objects, e.g., navigation through the interception of multiple radio signals from different locations and the calculation of time differentials. RFID took this an important step further, by adding the notion of active identification to the radio signals. One of the earliest applications was the "identify: friend or foe" (IFF) system of military aircraft, in which friendly aircraft were able to identify themselves through an automated response to radio signals.

One of the seminal papers on the use of RFID was written by Harry Stockman as far back as 1948: "Communication by Means of Reflected Power [1]." The paper discusses the potential of point-to-point communications enabled by radio, and mentions the following characteristics of such a system: high directivity, automatic pin-pointing in spite of atmospheric bending, elimination of interference fading, simple voice-transmitter design without tubes and circuits and power supplies, increased security, and, of course, a simplified means for identification and navigation.

Interestingly enough, it was just around that time, in 1947, that the first transistor was invented at Bell Labs. Ten years later, in 1958, the integration of a large number of tiny transistors into small electronic chips using semiconductor material was first made possible, i.e., the integrated circuit or IC (first developed by Texas Instruments and Intel). This landmark development ushered in today's information age. The rapid growth that ensued is aptly described by Gordon Moore [2], a phenomenon now known as Moore's Law. The combination of RFID and ICs is what now enables today's applications for automatic data capture and item-level tagging.

The 1960s saw the publication of a number of significant scholarly papers and inventions relating to RFID. It was also the beginning of commercial activities in the field. The first equipment using RFID to thwart theft was developed during this period. Electronic Article Surveillance (EAS) systems could only detect the presence or absence of a tag, but were highly effective in countering theft, and are considered to be the first widespread commercial use of RFID [3].

Innovation in RFID boomed in the 1970s through both publicly and privately funded research and development. For instance, the Los Alamos Scientific Laboratory in the United States and Sweden's Microwave Institute Foundation in Europe, were instrumental in advancing RFID technology. In 1975, the Los Alamos laboratory published some of their work in a critical paper on short-range radio-telemetry for electronic identification [4]. The nuclear industry was an early adopter of RFID during this period: both equipment and personnel were tagged in order to address concerns for security and safety. Countries like the United States began exploring the use of RFID in transport logistics.

The 1980s and 1990s introduced real-life commercial applications of RFID. In the United States, transport and access control were the main focus, while implants for cattle and other animals were secondary fields. In Europe, short-range systems were introduced for animals, and industrial applications such as toll roads emerged (e.g. in Italy, France, Spain, Portugal, and Norway) [3]. The first open highway toll collection systems were opened in the early 1990s (e.g. Oklahoma). It was during this time, as well, that microwave RFID tags (using higher frequencies but containing smaller antennas) were developed. These could be equipped with a single IC rather than an IC combined with an external inductively coupled transformer. This meant that RFID tags could be made smaller, cheaper, and more reliable than before, enabling the deployment of RFID on a much larger scale. Figure 4.1 provides a timeline of RFID development.

The purpose of RFID today is not vastly different from that first conceived in the middle of the last century, in that it relates to the use of radio waves for identification. However, with the rapid growth in personal computers, handheld devices, and the continuing increase in processing power, its potential applications are multiplying. At the turn of the century, many businesses were seized of the opportunity provided by RFID to streamline their supply chains, while others explored new application areas, such as environmental monitoring and food traceability. At the same time, RFID gained prominence in the media. In June 2003, news quickly spread of Wal-Mart's mandate requiring its top 100 suppliers to equip their shipping crates and pallets with RFID tags (and the Electronic Product Code or EPC by January 2005 [5, 6]). During that

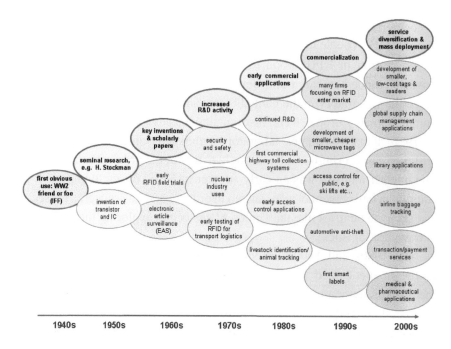

Fig. 4.1 RFID timeline.

same year, the United States Department of Defence also issued an RFID mandate. In 2008, Wal-Mart increased pressure on its suppliers to comply with its mandate, by charging suppliers a $2 fee for each pallet without an RFID tag shipped to its distribution center in Texas [7]. Adoption of the EPC standard (for data formats) by these large players will also have a significant impact. The network externalities [8] generated by these developments cannot be overlooked: the value of these systems will increase exponentially as the number of their users, and uses, continue to grow.

4.1.2 System Components

RFID systems consist of two main components: a transponder that carries data (typically known as a tag) and an interrogator that reads (and in some cases writes) that data (typically known as a reader). The tag is affixed on the item to be identified, and the reader can be handheld or mounted on a wall. RFID tags are seen as a future replacement for today's bar code (universal product code or UPC) that identifies almost all consumer goods today.

In most RFID systems, RFID readers are fitted with software, known as middleware, to enable them to forward data they receive to a controller. The controller could be a personal computer, a mobile phone, or even the global internet. RFID readers can read tags that are continuously shrinking in size. Many of the tags operating in the lower frequencies are no bigger than a grain of sand (i.e., less than 1/3 mm wide) and can be contained inside a glass or plastic module [9]. Hitachi made news when it developed one of the world's smallest tags, the μchip, which measures 0.4×0.4 mm in diameter. Since then, efforts at miniaturization and developments in nanotechnology continue to make tags smaller and cheaper. In 2007, Hitachi released a chip measuring only 0.05×0.05 mm, almost powder-like, and sixty times smaller than the μchip. However, this tag will not see the mass market for some time yet, as it lacks a radio frequency (RF) antenna (or inductive coil), and thus cannot be powered. Tags need to contain both data storage and transmission capabilities. Although data storage capabilities are similar in chips of both sizes, the smallest antenna developed thus far (for data transmission) is still around 80 times larger than in the new chip [10].

In a typical RFID system, the interrogator or reader will transmit a low-power signal to power the tag via its antenna (using radio frequency) or coil (using the magnetic field). The tag then signals back to the reader with data such as the tag's identifier (in the form of an alphanumeric code, such as the electronic product code). This identifier code can be associated with data held in a secured database, e.g., origin of the item, its date of production, and other relevant information. Most of the smaller low-cost tags are passive rather than active. Passive tags remain dormant until they are interrogated (and powered) by an interrogator within reading range (Figure 4.2). Active tags contain their own battery and can therefore send data to interrogators whenever they are within the reading range of the interrogator. They can also contain on-board electronics such as microprocessors and sensors. Active tags can be used in a much wider range of applications than passive tags. They also enable continuous data monitoring. Table 4.1 sets out the main differences between active and passive tags.

Managed RFID systems, made up of a set of readers and tags, can be closed or open. For instance, a retailer wanting to track movement of inventory may strategically locate readers around their stock room or warehouse, enabling them to read tagged items that are nearby. Data, such as location,

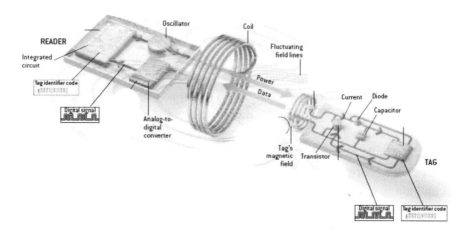

Fig. 4.2 Low-frequency RFID system.
Source: Adapted from Scientific American [11].

Table 4.1 Active and passive RFID tags.

	Active RFID tag	Passive RFID tag
Tag power source	Internal to tag	Energy is transferred from the reader via radio frequency
Presence of battery in tag	Yes	No
Availability of tag power	Continuous	Only when tag within field of reader
Monitoring of data input	Continuous	Only when interrogated by reader
Data/time stamp	Yes	No
Required signal strength from reader to tag	Low	High
Available signal strength from tag to reader	High	Low
Communication range	Long range	Short range
Data storage	Large (128 Kb)	Small (128 bytes)

Sources: Adapted from Auto-ID Center [12].

origination, date of expiration etc. could be transferred to a company intranet through a controller in a portable/mobile device, computer, or network of computers (Figure 4.3). A controller in an RFID system contains the intelligence of the system. Controllers network multiple readers and centrally process information received from them and associated tags. Depending on access restrictions and the desired functionality of the system, updated information in the company's database could also be transmitted over the public internet or over mobile networks (e.g. 3G). This information can also be made remotely accessible.

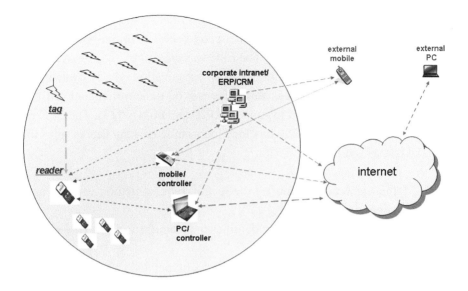

Fig. 4.3 Schematic of a managed RFID system.
Note: ERP stands for Enterprise Resource Planning and CRM for Customer Relationship Management. It is important to note that in a home environment, either a PC or a mobile handheld can act as the central controller for data storage, and for data forwarding either via the Internet or a mobile network (e.g. by SMS over a GSM network).

Tags come in many forms: active or passive, as mentioned above, but also read-only, read-write or WORM (write-once, read-many) format. Size and cost vary accordingly.

4.1.3 Standardization and Frequency Allocation

One of the main impediments to the mass deployment of RFID technology is the lack of global standardization. Global standards ensure interoperability between technologies and applications launched by different suppliers, industries, and countries. The absence of standards can hinder market development and delay the diffusion of innovations. In the fast-moving telecommunications industry, universal standards are all but indispensable for broader access to products/services, market expansion, reduction of complexity and cost, risk management, customer confidence, and the creation of economies of scale.

In the area of data formats for RFID tags, a popular industry standard has emerged, known as the electronic product code or EPC. Though EPC has emerged as a *de facto* standard in the United States and Europe, Japan has a competing standard, the *uCode*, developed by the Ubiquitous ID Center (see Figure 4.6 later in this section). The EPC was created by the Auto-ID Center and is now managed by EPCglobal — an organization wholly owned by the not-for-profit GS1. This organization focuses on the development of EPC as an industry-driven standard to support RFID technology [13]. The EPC is an alphanumeric code 64 to 256 bits long, consisting of a tag header, plus three sets of data (Figure 4.4):

- the *tag header* identifies the length, type, structure, version, and generation of EPC;
- the *EPC manager* code identifies the manufacturer, and the subsequent partitions;
- the *object class* identifies the product type; and
- the *serial number* identifies the specific instance (i.e. the particular item).

Readers can be programmed such that individual products, bearing EPCs, from a given manufacturer can be identified. This would enable the spotting of an item or batches of items in the supply chain for several different purposes, such as recalls or missing deliveries. In order to locate the relevant database, an "Object Naming Service" (ONS) converts the manufacturer's ID stored in the EPC into a web address. These ONS servers resemble the Internet's domain name system (DNS) but given the trillions of objects to be tagged, it is likely to be much larger than today's DNS. Software architecture for the EPC is set out in Figure 4.5.

Fig. 4.4 Electronic product code (EPC) layout.
Source: Sun Developer Network [14].

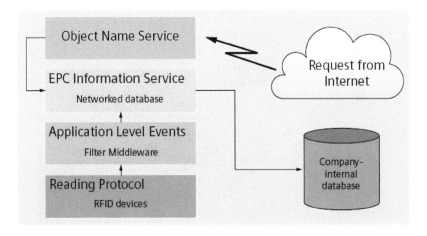

Fig. 4.5 EPC software architecture.
Source: Siemens [17].

Another technology that is emerging for item-level tagging and automatic data capture is the "ucode," developed by Japan's Ubiquitous ID Center [15]. The ucode is a system based on a 128-bit number, which can be attributed to a specific item or location. The code length can be extended to 128-bit units, e.g., 256 bit and 384 bit. Depending on the application, ucodes come in four different formats: printable codes (such as two-dimensional quick response or QR codes), passive RFID tags, active RFID tags, and active infrared (IR) tags. A device known as the "ubiquitous communicator" acts as the interrogator in this system. The architectural vision of Japan's ubiquitous ID system is set out in Figure 4.6. Unlike the EPC, the ucode, from its earliest inception did not limit itself to RFID tags as carriers. Another important difference is that the EPC network architecture relies on the internet for data transfer, whereas the ucode is compatible with different network types. The advantage of the EPC is that it was approved by the International Organization for Standardization (ISO) [16] in 2006, as a result, it is likely to emerge as the global industry standard for RFID tag formats.

Though data formats have generally fared well in terms of standardization, radio frequency use remains a significant technical and policy barrier to the mass global deployment of RFID systems. Frequency allocation for RFID varies greatly around the world, particularly in the ultra high frequency bands (UHF). This is further complicated by differing restrictions placed on

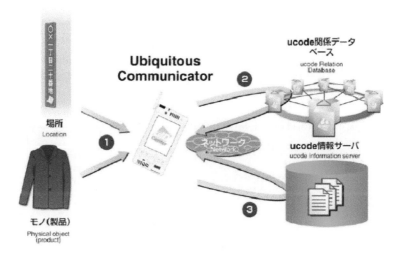

Fig. 4.6 Ubiquitous ID architecture based on ucode.
Source: Sakamura [19].

transmission power. Thus, readers are unable to process all kinds of tags [18]. Figure 4.7 sets out the different frequencies in use for RFID. At present, only the smaller, low-frequency (LF) tags have been mass produced. Both LF and high frequency (HF) systems can be used globally without the need for a license. This is not the case for UHF systems, which are required for applications requiring higher read ranges.

For instance, in North America, UHF systems ranging between 908 and 928 MHz, may be used without a license, but then transmission power is subject to restrictions. These systems cannot operate in Europe due to their interference with the mobile communications spectrum used for the Global System for Mobile Communications (GSM) and the spectrum reserved for military use (e.g. in France). In Europe, RFID equipment operates in the UHF band between 855 and 868 MHz. In Australia and New Zealand, 918–926 MHz is available for unlicensed use but once again there are restrictions on transmission power. In China and Japan, there is no regulation for the use of UHF. Each application for UHF in these countries needs a site license, for which an application can be made to local authorities, but is also subject to being revoked.

At the macro level, therefore, frequency harmonization is an essential hurdle to overcome before RFID tags become common place. At the micro (or firm) level, the choice of standard and frequency (low or high) is an important

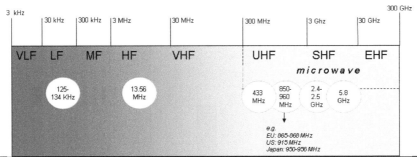

	Low Frequency (LF)	High Frequency (HF)	Ultra High Frequency (UHF)	Microwave 2.4–5.8 GHz
Typical read range	< 0.5 m	< 1 m	2–7 m (up to 100 m with 433 MHz)	1–2 m
Typical applications	Animal tracking, access control, POS applications	Smart cards & shelves, libraries, baggage tracking	Pallet/carton tracking, toll collection	Baggage tracking, toll collection

Fig. 4.7 Frequencies and read ranges for RFID tags.

Note: Read ranges listed above are indicative and can vary depending on the frequency, the type/size of reader and tag used (active/passive), and the environment.

one and should depend on the specific application under consideration. Higher frequencies provide greater read ranges but also require larger components. In addition to frequency use, another important consideration when building RFID systems is the selection of active or passive tags, as discussed above, as these affect read range and data monitoring capabilities.

4.1.4 Markets for RFID

Studies by market research firms and consultancies predict an explosion in the market for RFID. However, estimates vary greatly (Figure 4.8). IDTechEx estimates that in total, 2.16 billion tags will be sold in 2008 compared with 1.74 billion in 2007 and 1.02 billion in 2006. It puts the total RFID market value (including all hardware, systems, integration etc.) across all countries at USD 4.93 billion in 2007 and 5.29 billion in 2008 [20]. Projections released by the research firm Gartner are more conservative — it estimates that revenues from RFID technology will reach USD 1.2 billion in 2008 (up from 917.3 million in 2007) and USD 4.5 billion by 2012 [21].

Although market research firms may differ on sales figures, they agree on general growth trends. Moreover, some of the data is not directly

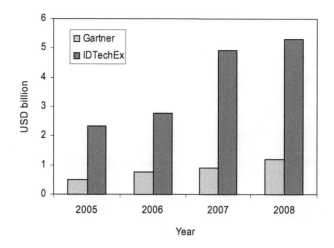

Fig. 4.8 Comparison of IDTechEx and Gartner estimates for worldwide sales of RFID.
Source: Adapted from IDTechEx and Gartner estimates [20, 21].

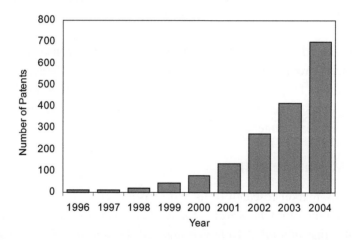

Fig. 4.9 Number of RFID patents per year.
Source: Adapted from Centredoc [22].

comparable: Gartner estimates, for instance, exclude contactless smart cards. Perhaps another indication of the interest in RFID by the manufacturing sector is the increase in the number of patents. According to the Centredoc RFID database [22], there was a 65 percent increase in the number of RFID patents in 2004 alone (Figure 4.9).

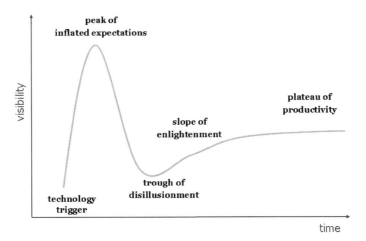

Fig. 4.10 The Hype cycle.

Source: Gartner Group [23].

It may be useful at this stage to recall the concept of the "hype cycle," first proposed by Gartner in 1995 (Figure 4.10). Gartner characterizes the excess of enthusiasm, or "hype" that can often accompany the introduction of new technologies, as a five-part cycle of varying length. It all begins with a "proof of concept" or "technology trigger," such as a major innovation or breakthrough in a product or service, generating significant press coverage and interest. Gartner describes the peak of hype that follows a "peak of inflated expectations" characterized by a frenzy of publicity. In this phase, some successful applications emerge but there are equally many failures and/or negative press coverage. As a result, this period is typically followed by a period of disappointment and disillusionment, as technologies fail to meet expectations. The press usually loses interest at this point, but businesses and innovators continue to experiment with the technology. When its benefits become more generally accepted, the "plateau of productivity" leads to a more stable product or service, be it for a mass or niche market.

In 2005, Gartner applied its concept of the hype cycle to the development and maturation of RFID technology. It concluded that the potential of RFID is being limited by the development of scattered applications under the same umbrella, and without any common thread. Nevertheless, the technology is predicted to reach maturity within a few years. The areas identified to be nearing maturity were asset-tracking and management (e.g. healthcare) and reusable tags (e.g. library materials). A few years on, in 2008, it would seem that RFID seems to have moved beyond the "trough of disillusionment" to an initial period of growth and productivity. But the market continues to be hindered by a lack of standardization and a dearth of end-to-end RFID solutions.

4.2   Applications and Functionalities

RFID technology can be applied in a variety of contexts, from transport to smart payments, from retail to healthcare, from military uses to farming. The real-time identification of individual items can benefit a wide range of industries, from banking to sports and entertainment. Gartner, for instance, has identified 99 different applications across 13 vertical markets. As performance continues to improve, and costs of tags and readers drop, tags will begin appearing on all kinds of devices and consumer goods, thereby multiplying functionalities and creating synergies across different industrial applications.

The better-known application types today are security and access control (e.g. securing corporate premises), transport and logistics (e.g. highway toll collection), supply-chain management (e.g. inventory control and re-stocking), and asset tracking (e.g. airline baggage tracking). RFID passes for holiday resorts (e.g. ski passes) and vehicle anti-theft mechanisms (i.e. immobilizers) are also increasingly common. Table 4.2 categorizes RFID applications into business, public sector, and consumer applications. Most examples cited are real-life commercial applications.

To grasp the full potential of RFID, it is important to consider not only the applications in use, but the various functionalities that the technology possesses.

4.2.1   Item-Level Tagging in Real-Time

First and foremost, RFID has the ability to identify individual items in real time. The existing bar code in common use only identifies batches or series of items. The advantage of the identifier on RFID tags is that they are able to identify each specific item or thing in a batch. For example, a handheld or wall-mounted RFID reader could determine whether a particular toothbrush is present among a large number of toothbrushes in a carton or a large number of cartons on a pallet without having to open each carton. RFID permits the identification of each individual toothbrush.

Table 4.2 RFID applications.

	Application type	Examples
Business applications	Transport & logistics	— Public transport (e.g., London's Oyster Card, UK)
	Security & access control	— Access to premises (e.g., EZPass)
	Mobile asset tracking & anti-theft	— Vehicle security (e.g., Texas Instruments, US)
		— livestock and animal tracking
	Supply chain management	— Inventory control (e.g., Wal-Mart, US; Metro Stores, Germany)
		— Anti-counterfeiting, for instance for pharmaceuticals
	Process manufacturing	— Automated assembly (e.g., Wells's Dairy US; Harley Davidson, US)
	Transaction/ Financial Services	— Taxi fare payments (e.g., Japan)
Public sector applications	E-government	— RFID in drivers licenses (under consideration, e.g., US)
	Defence & security	— Military defence tracking.(e.g. container tagging, US Department of Defence)
		— Passports (under consideration, e.g., US)
		— Currency (under consideration, e.g., US & EU)
	Healthcare	— Patient flow management (e.g., AHA Solutions, US)
		— Medical equipment tracking
	Environment	— Tracking waste & hazardous materials (e.g., Botek Vågsystem, Sweden & WasteTrax by AMCS, UK)
Consumer applications	Personal welfare and safety	— Tagging children at school or at amusement parks (e.g., Legoland, Denmark)
	Retail	— Tracking food while dining out for easy billing (e.g., Pintokona Sushi Restaurant, Japan)
	Entertainment and lifestyle	— RFID implants for access to VIP services (e.g., Baja Beach Club, Spain)

Item-level tagging not only refers to the identification of items, but also to the real-time visibility of their location and status. This form of tagging is particularly useful for supply-chain management processes such as inventory control. Information about the location and quantity of products can help firms to respond rapidly to customer demand. If a warehouse is running out of a product, RFID readers placed in so-called "smart shelves" could inform a central controller in real time. A command can then be issued to order additional stock, either from the distributor or the manufacturer. This streamlines ordering processes and saves not only transportation costs but customers from going to the competition for supply. Information gathered about product supply and demand can enhance procurement processes and bring costs down. The number of errors in invoicing and billing processes can also be reduced as the arrival of shipments are automatically tracked without having to verify delivery by opening packages or cartons. Eventually, as tags replace bar codes in the retail environment, customers would skip long queues at the check-out altogether, by walking past readers at store exits that automatically identify items in the customer's shopping trolley. Payment can then be made by whatever automatic means that are in place (e.g. credit card).

Tagging can also be used to ensure the authenticity of items and thus prevent counterfeiting. The pharmaceutical industry has already embraced RFID technology to prevent the distribution of unauthorized drugs. RFID systems could translate into substantial savings for that industry through fewer product recalls and less overstocking of drugs nearing expiration dates.

The ability to identify individual items does not only benefit large organizations. The tagging of things in households could help individuals better manage their personal belongings, digital gadgets, and household appliances. Combining tags with sensors can provide a simple solution for the transmission of sensor-related data. Commercial applications of this nature are not common, but are beginning to make their entry into the market. Universities and research labs are hard at work to design systems that would make day-to-day life more convenient, through "smart" gadgets powered by RFID. In 2006, for instance, Canada's Simon Fraser University designed a woman's purse that reviews its contents and alerts the owner if an important item is missing. The purse is equipped with an RFID battery-powered reader at its base, and RFID tags are affixed to designated items such as mobile phones, keys or wallet. The RFID reader is connected to a small screen with different patterns of light. Each pattern is associated with an item, and lights up if the item is missing. A similar prototype was created at Media Lab at the Massachusetts Institute of Technology (MIT) in 2004: the "build your own bag" (bYOB) project used RFID, light sensing, and Bluetooth to inform a bag's owner about the contents inside and near the bag [24].

The Dewey Decimal Classification (also called the Dewey Decimal System) used for the classification of books in libraries, designed in the 19th century, was an early attempt to constitute a system for the classification of individually identified things. Basic classes of things and knowledge were divided into ten divisions — each division could be sub-divided into ten sections, creating a total of 1,000 categories, which could be further sub-divided through the use of decimals, which could each be divided into ten sections, creating 1,000 categories, each with decimal sub-divisions [25]. During the next century, the internet was born, and currently runs a coding system based on 32 bit addresses, i.e., Internet Protocol Version 4 (IPv4). IPv4 allows the definition of about 4 billion distinct items. Work is underway to expand this to a 128-bit addressing system which could be used for item-level tagging. Such a system would permit the daily assignment of a trillion codes for a trillion years with room to spare [26].

4.2.2 Everywhere Computing at the Edges

Similar to the smart bags described above, RFID systems in combination with sensors can push computing to the edges of networks, and embed intelligence into everyday objects. This is more easily achieved with the use of active RFID tags that can initiate events and communicate data to readers at all times. The addition of sensors to RFID systems enables not only the collection of information about the identity of an item, but also about its state, e.g., temperature, velocity, the presence of bacteria, and so on. For instance, perishable food products could contain RFID tags equipped with sensors that determine whether an item has spoiled.

Furthermore, the addition of cheap low-power processing capability (such as that present in active tags) will slowly transform today's purely static objects into smart dynamic ones. This is the future vision of everywhere or "ubiquitous computing" which is likely to extend the scope and power of today's internet exponentially. Everyday objects could eventually become nodes in the network, processing information invisibly in the background, and enhancing the utility and function of networks and the data they carry. Ideally, such processing should not be visible to users and no additional complexity added to the user's experience.

Radio localization and virtual mapping: A number of different technologies have been explored for indoor geo-localization, including wireless local area networks (WLANs) based on the Institute of Electrical and Electronics Engineers (IEEE) standard 802.11 (e.g. Wi-Fi (Wireless Fidelity)), IR and Ultra Wideband (UWB). RFID, too, is now increasingly under consideration as technique for indoor localization. Because RFID tags are item-specific, they can be used to detect the absence or presence of things in space. Innovation in this field now enables readers to read multiple tags within their read range.

But the potential for RFID systems to precisely locate items in a given space has limitations. If a large number of tagged objects are present in a given space, it is possible to identify whether a particular item is in that space. However, it is much less straightforward to distinguish between different items or their locations. Passive RFID tags, in particular, are capable of self-identifying but not self-locating [27].

Although the Global Positioning System (GPS) provides an excellent architecture for determining location, it is costly and mainly unreliable in indoor environments. Efforts have been made to exploit the potential of RFID systems to triangulate the location of tags. In UHF systems using active tags, RFID readers use the electric field (far field) backscatter to power the tag and receive the tag's identifier. Since signal strengths vary depending on distance, it is possible to compare the signal strengths to estimate the distance between readers and various tags or sensors [28, 29], e.g., the SpotON concept. The different measurements can then be aggregated by a central server, which triangulates the position of a tagged object (Figure 4.11). In order to assist with location calibration, researchers at HKUST (Hong Kong University of Science and Technology), and MSU (Michigan State University) expanded

SpotON   **LANDMARC**

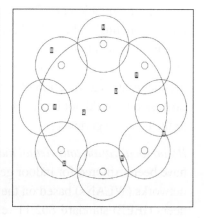

Fig. 4.11 SpotON and LANDMARC concepts for sensing location.
Source: Hightower *et al.* [28] (left) and Ni *et al.* [30] (right).

this idea by placing extra fixed "location reference" tags (instead of readers), and developed an algorithm to reflect the relationship between signal strength and power levels [30]. RFID systems have also been combined with Wi-Fi technology for real-time localization, particularly in the health care sector (see Section 4.5).

Stand-alone RFID systems have been combined with digital imaging in the Ferret prototype, developed at the University of Massachusetts [27]. Ferret is a system for locating nomadic objects tagged with RFID and displaying them to a user in real-time using a handheld video camera with a display and embedded RFID reader. This is an interesting concept, but the utility and convenience of scanning a room with a camera each time the location of an object is sought may require further assessment.

Although a number of pilots and experiment have been conducted on RFID localization, much work remains to be done. The establishment of this facility is likely to find widespread application and thus prove a significant business opportunity. It can introduce context-awareness to the future internet of things: RFID readers and tags, in combination with sensors, would then form part of a necessary bridge connecting the real and virtual worlds, both in outdoor and indoor environments.

4.2.3 Technical Limitations

Despite the valuable functionalities that RFID provides, the technology suffers from a number of important technical restrictions and limitations, which require further study and innovation.

One of the main challenges ahead is the seamless integration of sensors and actuators with RFID systems. Although the combination of identification technologies and sensors is being touted as a vital enabler to the future internet of things, much work remains to be done in this area, both in terms of systems integration and standardization. Communication between tags and sensors, e.g., using mesh networking, is also an area that requires further study.

Another concern is the possibility of false-negative readings, that is to say situations in which the tag is not detected by a reader in spite of its proximity. Similarly, false-positive readings are also possible. For example, due to the orientation of the tag and reader, the tag's energy can sometimes be absorbed by the tag's own antenna [31]. In addition, materials such as metal and water (e.g. like a human body which is mostly water) can interfere with the detection of tags [32]. Another key limitation of RFID systems is their sensitivity to environmental conditions. And for active tags, battery life is limited: techniques such as energy harvesting and scavenging need to be further developed.

RFID systems suffer from technical constraints related to collisions between readers and interference from other kinds of radio-frequency emissions. Equally, maximum read ranges and power transmission restrictions can limit the functionality and suitability of RFID systems for certain applications.

Systems that use UHF and microwave frequencies suffer from "ripples" in the coverage area of a reader, given the use of dipole antennas and the relatively short wavelengths. Signal strength is therefore not always uniform throughout a read zone. Unlike LF and HF systems, it is possible for a read zone to have areas where signal strength is as low as zero, giving rise to "antenna nulls" or blind spots [33]. These blind spots must be taken into account when designing UHF RFID systems.

As the number of RFID tags increases, scalability becomes an important concern. The processing of large amounts of data collected by a multitude of tags poses a challenge to data processing and systems management. Furthermore, technical solutions for data security and privacy need to be put into place before mass adoption becomes acceptable.

4.3   RFID in Health Care

Information and communication technologies have long been seen as vital to modern health care information, management, and delivery. Computing and electronic health records (EHRs) are the commonplace in developed countries, and network technology is being increasingly used to deliver health care to hard-to-reach areas in both the developed and developing world (i.e. telemedicine).

Many analysts believe that the healthcare and pharmaceutical industries will be early adopters of RFID technology on a large scale. In the United States, the Food and Drug Administration (FDA) endorsed the use of RFID to track pharmaceuticals and has called it the most promising technology for implementing electronic track and trace in the pharmaceutical supply chain. It has also approved the human RFID implant known as "Verichip," designed by the firm Applied Digital Solutions [34].

Table 4.3 Forecasts for RFID in the health and pharmaceutical industries.

	2007	2012	2017	2022
RFID tags (millions) on hospital assets	2	98	190	320
RFID tags (millions) on samples	1	8	30	40
RFID tags (millions) on drugs	5	246	1,500	6,380
Total RFID tags (millions)	9	350	1,710	6,740
Locations with RFID readers	110	2,770	11,900	40,600
Total number of RFID readers	180	12,600	70,200	208,000

Source: BRIDGE Project (European Commission) [37].

Logica and GS1 estimate that the number of RFID tags used in healthcare will grow dramatically from just 9 million to 1.7 billion by 2017 (Table 4.3). IDTechEx estimates that the market for RFID tags and systems in healthcare and pharmaceuticals will rise from USD 85 million in 2007 to USD 2 billion in 2017. According to the market research firm, the main growth areas are item-level tagging for pharmaceuticals and location systems for staff, patients, and assets [20]. A timeline for RFID healthcare applications is set out in Table 4.4.

Analysts predict that current market and technological conditions are optimal for the deployment of RFID technology in the health care sector [35]. The vast majority of such RFID applications have thus far been deployed in hospital

Table 4.4 Timeline of RFID application development in the healthcare industry.

Up to 2004	2005–2010	2011 onwards
— Error prevention of products (e.g., blood ID, correct baby/mother match)	— Error prevention now includes autorejection of incorrect parts or luer connections	— Location of visitors, visitor alarms, & virtual queuing
— Patient tagging for error prevention	— Location of staff now coupled with staff alarms/tags that record incidents	— Patient compliance monitoring/ prompting (intake of medication)
— Location of staff and staff alarms	— Speedier asset location and more accurate stocktaking	— Track and trace of most pharmaceuticals, consumables and assets
— Location of assets	— Prevention of theft	
	— Cost control	
	— Recording of procedures (e.g., for use in lawsuits)	
	— Drug trials compliance — monitoring & prompting	
	— Behavioral studies in order to optimise operations	
	— Pharmaceutical anti-counterfeiting	

Source: Adapted from IDTechEx [41].

settings, e.g., asset management and tracking, patient and staff monitoring, and bedside error prevention. These areas are discussed in more detail below. RFID for the delivery of care in the home is considered in subsequent chapters.

4.3.1   Asset Tracking

Asset management (e.g. equipment and pharmaceuticals) is currently the largest field of application of RFID in healthcare. In surgical settings, every-day procedures can often require a large number of instruments (typically on trays) as well as devices and equipment, such as intravenous poles, infusion pumps, operating table accessories, monitoring cables, bronchoscopes, endoscopes etc. [36]. It is as vital for all trays of instruments to be present at the onset of surgery, as it is for them to be sterilized and in good working order.

With the use of RFID tags, smart inventory cabinets can be deployed in Operating Rooms (ORs). Required equipment can be easily traced wherever located in the hospital and be brought speedily to the operating room. Information about the location of tagged equipment would be transmitted to strategically placed readers throughout the premises and passed along to a centralized database. This information could then be queried by staff as required, through a simple user interface (e.g. a web browser) available on hospital PCs. Information about the tagged items can range from general presence in a room (e.g. when equipment passes a particular doorway) to real-time location status (e.g. through more advanced triangulation methods) [38]. The most common real-time location services in healthcare are:

(a) *Zonal RFID systems*:
   In these systems, readers are installed throughout the hospital to ensure that tags on items are always within the range of at least one reader.
(b) *WLAN-enabled RFID systems*:
   These systems use a combination of IEEE 802.11 Wi-Fi technology and active RFID tags. Tags are read by Wi-Fi routers that are also RFID readers [39, 40].

Most asset management in hospitals relies on passive RFID tags. For more precise fixing of location, active tags are preferable, as they are able to send information about their status spontaneously and do not require activation by a reader. System integrators interested in the pharmaceutical market tend to favor RFID systems operating in the HF range, as they are more forgiving of materials such as water and metal, have a more predictable read range, and can be used with smaller tags for integration with smaller items (e.g. individual pills [41]).

As mentioned earlier in this section, RFID is increasingly being used to thwart drug counterfeiting — a nefarious practice that results in significant financial loss to pharmaceutical companies and poses an important risk to patient safety around the world. RFID tags (with their unique identifiers) can be affixed to every case, pallet and package of medication, tracking it from the laboratory to the pharmacy or hospital. Such visibility of the supply chain would help ensure that medication reaching patients is authentic and has not been interfered with.

Another important application of RFID for asset management is the prevention of medical equipment or instruments being left inside a patient's body, which happens roughly once in every 10,000 surgeries [42, 43]. Among the most common items left behind are sponges. Medical staff must be procedurally required to count sponges before and after a surgery and errors are known to occur. RFID tags embedded in sponges and other objects could help detect such problems. A 2006 study of eight different surgeries conducted by the Stanford School of Medicine found that the RFID reader (in the shape of a wand) was almost instantaneously able to detect every sponge inside a patient's body without any false reports [44]. Though found cumbersome, the medical staff was pleased with the accuracy of the system.

In general, the benefits of tracking assets in a hospital setting are tremendous: avoidance of delays and errors in surgery, promotion of patient safety, optimization of business processes (e.g. medication dispensing) and workflow, introduction of more effective and accurate inventory systems, and avoidance of theft [45].

4.3.2 Bedside and Surgical Error Prevention

RFID can serve to reduce errors before, during, and after surgery through the electronic provision of health records, and the identification of important items such as blood samples, organs, and prosthetics. In the peri-operative setting, patient safety issues have been broken down into the following three key areas:

1. *Right patient, wrong treatment*: e.g., drug allergies and incompatibilities, wrong instruments made available, surgeon not available, surgery on wrong portion of the anatomy.
2. *Right patient, no treatment*: e.g., lack of specialized resources required for a particular treatment (or consequences of surgery);
3. *Unknown patient, undetermined resource* [36]: e.g., transfusion of wrong blood type.

In the case of blood transfusion, RFID tags (e.g., self-adhesive labels) can be affixed to easily identify the origin, type, and destination of a particular bag of blood and ensure it is delivered to the right patient at the right time. RFID readers (typically handheld) read the unique identifier on each bag of blood, together with a patient's wristband, and thus greatly reduce the chances of error.

RFID can provide rapid and accurate access to information about a patient's current medication and possible incompatibilities, notably in ORs and intensive care units. Current bedside checks are often done through a manual reading of documentation and are thus subject to human error. In emergency situations, the rate of error is likely to be exacerbated due to factors such as stress, distraction, and lack of resources. In both types of bedside checks, traditional bar codes require a flat surface and "line-of-sight" information scanning, which pose a particular problem when clothing and surgical material are present. RFID presents an interesting solution to this problem, as it provides simpler contactless solutions for identification.

4.3.3   Patient and Patient-Flow Monitoring

The use of RFID for patient monitoring is increasingly widespread. Active tags are preferred for this purpose, as are read/write tags, due to the need for real-time updates to medical information present on the tags. The most popular method for identifying patients is a wristband with a bar code, or in the case of RFID, a tag with electronic identifier. This wristband can be worn from hospital admission to discharge. This can be very crucial in cases where a patient is unconscious and cannot self-identify. With active tags, location information can be stored and automatically transmitted at regular intervals to readers. A central controller compiles the information in a database. The location of hospital staff (particularly in the case of emergency surgery) can also be determined using RFID tags, although this raises critical social issues such as the degree of employee monitoring and the protection of privacy.

Other identification systems with longer shelf lives are under study, e.g., RFID tags fixed on a tooth, over a protective resin [35] or RFID tags that are implantable under the skin. The most well-known implantable tag is made by Applied Digital Solutions. The latter has developed a patient identification system known as VeriMed. Most of their implantable tags, however, are in the early stages of development: they are being tested in hospital pilot programmes or at major worldwide conferences. The two-year pilot programme run by New Jersey's Horizon Blue Cross Blue Shield, in conjunction with Hackensack University Medical Center, is a good example. In this trial, participating patients with chronic illnesses received microchips inserted just under the skin, in their right arm. The unique identifier would refer to information (e.g. family contacts, lab test results, prescription records, patient's condition) located in a central database (known as the VeriMed Patient Registry) that would be accessible only by the Center's doctors [46]. The combination of RFID tags with various types of medical sensors, such as glucose sensors, is an important area for further medical research and innovation. VeriChip, for instance, released its own white paper on RFID glucose-sensors for diabetics at the end of 2007 [47].

RFID has the potential to help monitor the intake of medication by patients, through smart shelves and smart medicine bottles. But RFID tags can also be integrated into each pill. Such ingestible tags would be made of biologically inactive metals, such as gold, and would be easily destroyed by interactions within the body, such as interaction with hydrochloric acid [35]. Such systems could also prove vital in monitoring the intake of medication in homecare environments.

In sum, RFID cannot only track the flow of patients, staff and visitors throughout the hospital, but can also assist with the monitoring of individual patients in both hospital and nonhospital settings.

4.3.4 Limitations of RFID in the Healthcare Context

Although RFID holds great promise for enhancing the accuracy and cost-effectiveness of health care processes, much work remains to be done.

As discussed above, standardization and interoperability pose a significant challenge, as do the reliability and robustness of RFID tags, readers, and middleware. The possibility of interference with medical equipment, such as pacemakers and heart monitors, still stand in the way of full integration of RFID in hospital and homecare settings [45]. Equally, tags are not yet small enough for applications such as the tagging of small surgical instruments or sponges. Ingestible RFID tags, too, are not in a sufficiently advanced stage of development and may benefit from the development of entirely new materials [35].

One of the main barriers to RFID adoption in the healthcare industry is the fear of system failure. It is imperative for managed health information systems to be available and functional, particularly in critical emergency situations. For instance, the display of accurate information about the identity of the patient, allergies to medication, the reason for admission etc, are crucial during surgery and emergency procedures. The need for an information technology (IT) expert to intervene in case of a system malfunction can complicate surgery [38], and even lead to life-threatening situations. Health professionals are concerned about excessive dependence on information technology and the necessity for staff to acquire more than basic skills in information technology. For these reasons, system complexity must be kept at a minimum, and the user interfaces simple and easy to use.

Another important issue is the link between data stored in an RFID-enabled system and data stored in more widespread EHR systems. Data formats must be made interoperable. Mechanisms for secure data transmission must also be put into place, in order to exploit the full potential of RFID use in this sector.

4.4   Connecting Sensors

The main benefits of RFID for healthcare stem from its use as an identifier and as a transmission protocol. In order to maximize its utility for monitoring health, it is best used in combination with sensors, such as biomedical sensors, temperature sensors, blood pressure sensors, and even motion sensors to detect movement. RFID systems for health care could use tag-only systems for location and identification purposes, and sensor-enabled tags to facilitate data flow from medical devices and sensors.

The use of sensors has been expanding in every domain. They will soon grow from occasional use to universality. And their employment in tomorrow's world will become generalized. Already, they are found as useful in the home as in offices and facilities. Their power lies in the ability to complement the senses, in gathering pointed information about the surrounding circumstances and responding to them. While they have been in use for some time, recent research and innovation in this area has led to the development of smaller and smaller sensors that are more accurate, secure, and sophisticated for a growing range of applications across industries. Although there are some RFID tags on the market that are equipped with sensors, applications developed are fragmented and often limited. The potential benefits of combining sensor and RFID technology, though great, have yet to be fully realized.

4.4.1 Quality and Quantity

Simply put, a sensor is a device that responds to a physical stimulus by transmitting an analog or digital signal that can be interpreted by humans and machines. Many simple sensors have existed for some time, such as the well-known mercury thermometer. It measures the temperature of its environment on a continuous basis, because the material it contains changes its characteristics (i.e. expands) as temperature rises. This simple sensor is calibrated in such a way that temperature can be determined by the physical markings along it. Other examples of sensors in everyday life include touch-sensitive elevator buttons or lamps that respond to the sound of clapping [48].

Sensors consist of two important components: a sensing component, which detects the physical stimulus, and a communication component, which transmits the data gathered. Sensors need appropriate calibration to function properly, so that a standard output is defined for mass use.

Today, sensors are used across industries and product segments, from healthcare and robotics to automobiles and construction. They are typically classified according to the parameter or stimulus they are intended to measure, for instance:

- mechanical sensors (that detect, e.g., position, acceleration, force, humidity, pressure, motion, or shape);
- thermal sensors (that detect, e.g., temperature, thermal conductivity, or specific heat);
- electrical sensors (that detect, e.g., charge, current, voltage, or conductivity);

- magnetic sensors (that detect, e.g., changes to the magnetic field, or permeability);
- optical sensors (that detect, e.g., refractive index, or absorption);
- acoustic sensors (that detect, e.g., wave amplitude, wave velocity, or spectrum);
- chemical sensors (that detect, e.g., humidity, ion, gas concentration);
- biological sensors (that detect, e.g., toxicity, presence of biological organisms) [25].

Different types of sensors can detect different types of stimuli, e.g., humidity can be detected not only by mechanical sensors but also by electrical sensors. Both qualitative data and quantitative data can be generated, e.g., the amount of humidity can be detected, but also the absence or presence of impurities.

Although the early mercury thermometer was based on a nonelectrical, analog sensor, most sensors today have the capacity to emit an electrical signal. They convert the parameters they measure into changes in electrical attributes, i.e., resistance, capacity or inductance. These in turn affect the voltage or current that can generate the output or electrical signal. For communications with computers and machines, an embedded analog to digital converter (ADC) is typically present to convert the analog data into a form that can be understood and manipulated by a microcontroller.

The choice of sensor will depend on the environment in which it is used and the particular application in question. There a number of factors to consider when choosing the type of sensor required, e.g., accuracy, cost, and sensitivity to the environment. Sensors also have technical constraints that must be borne in mind, such as power, size, storage, and processing capability.

It can be said that sensors drive decision-making processes, be these human or machine. They are programmed to detect changes in environmental conditions and thus enable systems to respond more suitably. It is in this role that they can be seen as kingpins of such systems. When combined with actuators, they can take programmed action, such as varying the room temperature as required. With the growing interest in human–computer interaction, the role of sensors is likely to become more and more vital. Already scientists are finding ways to combine sensors with the ubiquitous mobile phone, incorporating glucose meters, and breathalyzers. Making sensors mobile gives them more autonomy and collaborative potential with other objects [49]. For context-based digital applications and environments to become a reality, such as home robotics, computers will have to rely increasingly on data detected by sensors.

4.4.2 Sensors and Radio

In order to serve their purpose, sensors must transmit data about the changes they detect to a central entity. This is the second necessary function that they must fulfil. Before a sensor can transmit data, that data must first be converted to a digital format via the ADC to make it machine readable. Next, its data must undergo a minimum of digital signal processing (DSP) in order to prepare it for analysis by a controller, e.g., in the case of a digital thermometer, translating a digital signal into a more precise average temperature.

Sensor data in this format can be easily accessed by a central controller via physical cable or a client (such as a PC or mobile phone). However, data can also be transmitted over the air interface using radio technology, which enhances both the availability and impact of sensor technology.

There are choices for transmitting sensor data via radio, e.g., Zigbee. But the platform developed during the course of this research uses RFID as the transmission protocol. The use of a simple RFID active tag combined with a sensor can dispense with the need for multi-layer protocol stacks between sensor and client. When a sensor is activated, it can in turn activate the tag, which can send its unique identifier to the nearest RFID reader. As a result, the reader is aware of a change of status. This change of status is then communicated to a central controller for processing and further action.

4.4.3 Shrink, Think, and Network

Sensors today are under intense scrutiny and development the world over. Innovation is particularly centered upon reduction in size. Both RFID tags and sensors are shrinking in line with developments in nanotechnology, e.g., nano-materials and nano-processors. Reducing sensor size will mean that they may become even more ubiquitous and embedded in the environment around us. Their presence will also be less obtrusive, enabling everyday objects to become "sensitive" and facilitating environmental monitoring on a larger scale (which is of particular importance for watch and warn systems related to environmental or biomedical hazards/disasters).

Sensors may soon become invisible to the eye, and at the same time, acquire more complex sensing and processing power. The disappearing processor will soon be joined by the disappearing sensor [50]. Advanced research laboratories such as University of California in Berkeley and funded by the United States Defense Advanced Research Projects Agency (DARPA) refer to "smart dust," implying that sensors could eventually shrink to the size of a particle of dust. This may be some decades away, but today the five millimeter sensor is already a significant achievement. Scientists are also considering different ways to manage power for sensors, and in particular wireless sensors, through energy harvesting or solar cells [51].

Not only are they shrinking in size, but sensors are becoming increasingly independent from human intervention. They can be programmed to take autonomous action to normalize and diffuse an environmental or object-specific crisis.

When networked, sensors can be made to think and act collectively. Each sensor "node" in a wireless sensor network is a small, low-power device, which normally includes the sensor itself, together with power-supply, data storage, microprocessors, low-power radio, ADCs, data transceivers, and the controllers that tie all components together. Applications for these wireless sensor networks are wide-ranging and include indoor location sensor systems, environmental monitoring, process automation, wildlife habitat monitoring, public safety, emergency services, environmental watch, and warn systems [25].

As a result, there is much interest in research and innovation in this area. One of the more promising developments is the possibility for sensor nodes to self-organize. Self-organizing networks (SON) can gather and process information by whichever node is available rather than using a pre-determined order of nodes. SON systems are organized without any external or dedicated central control entity and can better respond in real-time to quality of service demands and unpredicted changes in the network [52].

4.5   RFID and Its Expanding Orbit

RFID is one of the more promising technologies today in the field of automatic identification and short-range data transmission. However, it must be evaluated in its expanding context. This section considers the connectivity of RFID systems and new developments in tagging and data capture technologies. It also discusses the vision behind a future network of tagged things.

4.5.1 Networking RFID

The selection of a networking platform is a choice for industry, very much based on cost, standardization efforts, and industry partnerships. To connect various devices (PC, mobile phones, personal digital assistants or PDAs) to a central internet access point, WLAN has often been favored due to its availability, ease of installation and use of unlicensed spectrum. It is a much simpler solution than rewiring a space with, for instance, an Ethernet network. Not surprisingly, the use of RFID in combination with enterprise WLAN networks is growing, as it does not require the deployment of additional network infrastructure. Both WLAN and RFID are increasingly being examined in the development of localization services. WLAN-enabled RFID equipment is now available commercially [53]. RFID readers and tags are increasingly embedded in mobile phones (e.g. such as the Nokia 5140).

Zigbee, a wireless protocol based on IEEE 802.15.4, is among the more promising technologies for enabling appliances and devices to communicate wirelessly with each other, regardless of manufacturer. Zigbee functions over a range of ten meters and is largely used for wireless personal area networks (WPANs). It can provide a low-speed low-power network platform for RFID devices to communicate with each other and other devices. Zigbee has been used with RFID to provide patient location systems in hospitals, coal mining safety and fire department safety systems. Like WLAN, Zigbee's key advantage for home networking is that it is an international standard backed by a strong industry alliance [54]. However, it is not as widespread as WLAN and is still relatively costly. It remains to be seen whether Zigbee will emerge as an industry leader for health and home care environments over the near term.

Bluetooth [55] is yet another promising short-range wireless protocol, though it is seen as a replacement for cables and wiring rather than a networking technology *per se*. Bluetooth can connect and enable data exchange, over a range of 10–100 meters, between different devices such as PDAs, mobile phones, laptops, PCs, printers, headsets, printers, digital cameras, and video game consoles.

The short-range exchange of data can also be effected over IR. Infrared [56] data transmission can be used to connect computer peripherals and accessories and it, too, is particularly well-suited to personal area networks. Today, it is most commonly used for remote controls that command appliances.

Personal area networks are seen as pivotal to enhancing quality of life, and can help enable smart, responsive and portable environments. Networking RFID readers and tags can be an important element for making environments context-aware and responsive. It is not surprising that personal networking is an important priority of the research agenda of the European Commission. In this context, the MAGNET (My Personal Adaptive Global Net) project is worth mentioning, as it centers on the user and supports unobtrusive and portable personal networks. A Personal Network (PN) takes the notion of the PAN a step further by removing the barrier posed by geographical boundaries [57]. The PN vision focuses on the creation of virtual personal environments that can provide true mobility, whether in the home or out of the home, on the road or in the office (Figure 4.12).

4.5.2   Tagging and Wireless Data Capture

There are a number of technologies that will serve to widen the scope and import of RFID systems. In particular, sensors and wireless sensor networks, as discussed above, will play a vital role. But there are also new developments in tagging technologies that should be closely watched.

Innovation in surface acoustic waves (SAW), for instance, could dispense with the use of a processor altogether, by using "chipless" tags. Without the need for a processor, tags can be produced at a much lower cost. Although acoustic wave devices have been in commercial use for more than 60 years in the telecommunication industry (e.g. as filters in mobile phones and base stations), studies are being conducted on their potential for use in transmitting identity. SAW converts low-power microwave radio frequency signals into ultrasonic acoustic signals through the use of a piezoelectric crystalline material in the transponder. Variations in the reflected signal can be used to provide a unique identifier. SAW can be used in conjunction with RFID: a SAW RFID device might be a coded IDT (interdigital transducer) coupled with an antenna. An interrogating radio signal is received, generating an acoustic wave. The acoustic wave is partially reflected by each reflector, and converted back to an electrical signal that is then transmitted [59].

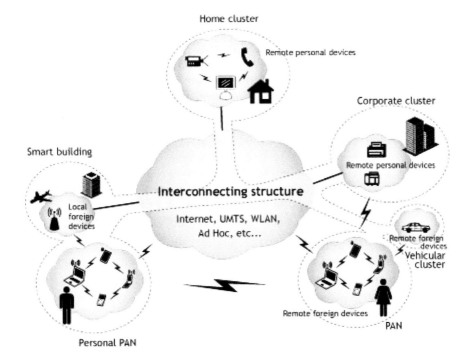

Fig. 4.12 MAGNET: The vision of the personal network.
Note: UMTS stands for Universal Mobile Telecommunication Systems.
Source: MAGNET [58].

The use of UWB is also being explored for real-time location systems. UWB is a wireless radio technology that was originally developed for secure military communications and radar, and was initially known as pulse radio. Its advantage is that it is able to transmit large amounts of data over a wide spectrum of frequency bands with very low power for a short distance. It can be used for short-range high-bandwidth communications. It is based on spreading the spectrum (over approx. 500 MHz) in order to share spectrum between users. UWB could help RFID overcome its shortcomings in harsh or cluttered environments, extend the range of tag interrogation, enhance anti-collision measures in multi-tag environments, reduce the number of antennas required for coverage, and provide better precision (down to 5 cm). UWB RFID tags are also less prone to inaccurate or false positive read results. At this time, UWB active tags are still fairly expensive (typically €40 per tag) [60]. Research and development is ongoing on the development of passive and semi-active UWB tags, which may further facilitate monitoring of the status of objects in real-time [61]. UWB RFID tags are of particular interest for the health care sector, given that they provide a greater degree of precision and do not interfere with electronic equipment used in hospitals [62].

Other technologies in their nascent stages include the use of DNA fragments in tags and optical tags. Tags in the future will be increasingly ubiquitous, even ingestible and biodegradable, and are likely to form an integral part of the objects in our environment.

4.5.3   A Future Internet of Things?

RFID is viewed as an important building block for the future of "internet of things [25]." The ability to automatically identify individual objects in our environment has the potential to revolutionize the internet of today. The internet has already come a long way from its beginnings as a communication network between some university centers, to a network used by over a billion people the world over. It connects anyone, anytime, anywhere. But technologies like RFID can extend the reach of the internet by connecting not only people and machines, but things as well. One would then achieve connectivity for anyone and *anything*, anytime and anywhere. Humans seem set to become a numerical minority on the internet, as devices, machines, robots, and designated items in households begin to appear on it. This view is based on the nature of emerging technologies and research and development activity. RFID, its associated identifiers, and sensor technologies will play an important part. The crucial capability is that of collecting real-time raw data about things, enhanced by status and sensory information. Connecting things in the new environment promises great market potential and can be a significant growth area. Smart homes and smart spaces full of all manner of connectivity will become the most natural thing in the world. The sense of "novelty" will disappear.

As Mark Weiser, the late Xerox computer scientist, aptly put it: "The most profound technologies are those that disappear. They weave themselves into the fabric of everyday life until they are indistinguishable from it [63]." In the future, perhaps all our spaces will be smart and intuitive, our communications ambient, and our networks invisible.

The building of this future, however, this brave new world [64], calls for a multi-disciplinary approach and must rest on a superlative understanding of social implications.

5

At Home with Age: Smart Wireless Living for the Elderly

Old age is like everthing else. To make a success of it, you've got to start young.

— Fred Astaire

This chapter outlines the AGE@HOME platform, which has been formed on the basis of the preceding research and analysis. It begins with an overview of developments toward the digital home and the main challenges encountered. It then presents the background, objectives, and overview of AGE@HOME before introducing the decision-support model that will be described graphically in Chapter 6.

5.1 Networking the Home

The prevalence of digital media, together with increased technical and industrial convergence, is in the process of revolutionizing the home environment. In this context, wireless connectivity is seen as a crucial element in creating convenient and smart living spaces to enhance the quality of life of residents at home and away.

The deployment of fully networked homes will most likely be an incremental process [77], at least in the early stages, rather than a revolutionary one. In other words, it is unlikely that traditional homes will be rapidly uprooted to be replaced by newer digital homes or that, in the near term, housing developers will build networked digital homes on a mass scale in most countries. For this reason, industry should emphasize the integration of current disparate home network systems, devices and appliances, both traditional and nontraditional, high-tech and low-tech. This will enable residents to incrementally transform their home spaces into networked and security-enabled living, entertainment and communication hubs. To this end, platforms that are interoperable and standardized will limit barriers to use, for they will be more convenient, afford-able and widely available. The use of wireless networks is seen as preferable as they provide the ability to interconnect various devices in the home with ease, while the "re-wiring" of an entire household is typically prohibitive in terms of cost and inconvenience.

Some efforts are already under way to further automate and network the home beyond the provision of wireless access to internet and broadcast media. In particular, video and music entertainment services are being integrated on so-called centralized "media centers" run by home computers. However, most networked applications for the home are being implemented as standalone systems and this poses a great challenge for further development and adoption.

Furthermore, most digital homes of the future will place increasing finan-cial burdens on residents and homeowners. Users will also face increasingly onerous tasks related to the management and administration of more and more complex systems. Thus, to stimulate adoption, strong incentives must be pro-vided: simplicity and affordability are vital, but so, too is the targeting of specific user needs.

5.1.1   Current Systems and Limitations

A number of home media, entertainment, communications networks, and secu-rity systems are currently available for use in the home. These include:

- mobile and fixed voice communications;
- broadband internet access;
- messaging applications;
- alarm systems;
- entertainment systems, including networked entertainment sys-tems (e.g. IPTV, gaming);
- home data storage (i.e. local and internet server capabilities);
- home automation (e.g. RFID car locks and garage door openers).

Some of these systems have recently begun to converge. For example, digital cameras allow the simple creation of slide shows and films for home use. These can be viewed on televisions or home computers. They can also be shared with others via e-mail or through shared web site access. Terrestrially transmitted digital TV programs can now be watched on personal computers.

Using a set-top box, users can watch programming delivered over a cable connection to the home (Cable TV or IPTV) or via satellite. As mentioned above, home computers are being designed or adapted to act as home entertainment hubs or media centers that provide film and television entertainment, music, and broadband internet access all in one.

Nevertheless, most home appliances or systems that are equipped with a communications component are provided as standalone systems today. Intruder alarm systems have always been fairly independent — they have their own set of connected wired or wireless sensors and embedded means of communication. A variety of systems on the market also alert for abnormal ambient temperatures, flood or fire: like many home or perimeter security systems, alerts can be sent to pre-programed telephone numbers or alarm call centers. Entirely different distributors typically sell networked kitchen appliances (such as internet-enabled refrigerators). Similarly, home automation systems (e.g. automatic garage doors, lighting control, adjustable video surveillance etc.) are largely standalone systems and are often sold to hobbyists rather than to a wider market.

With respect to other types of home monitoring systems (e.g. environmental or resident monitoring), these are beginning to appear on the market, but adoption is slow. Service providers are typically different from those providing the services or products mentioned above, and can range from large network operators to smaller start-ups. Among commercial offerings are British Telecom's (BT) Home Monitor service [1] and Manodo's Home Arena [2]. These systems combine energy consumption monitoring with intruder alarms and emergency alarms. However, none emphasize the monitoring of unusual activity on the part of a resident, and as such cannot effectively function as independent living platforms. The BT service, for instance, includes intruder protection through deterrence (i.e. a warning siren), a panic alarm for personal protection, and real-time alerts (SMS or fixed line phone). The Manodo system (Figure 5.1) takes a different approach by focusing on metering and the analysis of household energy consumption. One of its add-on capabilities is a camera application at the front door, enabling residents to view visitors at their front door (through an MMS sent to their mobiles) before choosing to remotely open the door if they wish.

Even if home automation technologies have been available since the 1970s, they have not served to improve home environments dramatically and new

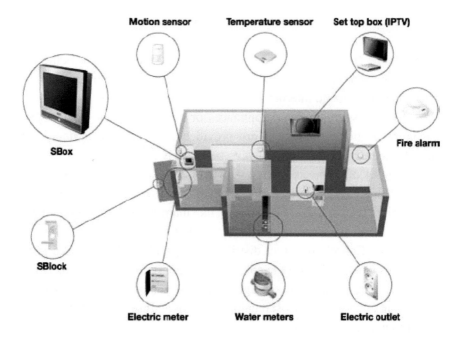

Fig. 5.1 Manodo home arena platform.

Source: Manodo [3].

developments remain a far cry from the vision of the fully networked digital home. One of the greatest challenges is that users do not consider the overall benefit of these systems to be justified by their cost. Furthermore, many existing systems are inflexible: they are not open to changes or variations in lifestyles and can often require an upgrading of the home network infrastructure. Finally, many of the early systems have not taken sufficient note of user expectations or requirements [4].

Little effort has been made so far to bring disparate home technologies and systems under one umbrella. It seems, therefore, that the time is now ripe to actively integrate functions being delivered by different providers on various platforms. Ideally, this is best undertaken through an understanding of the specific needs of residents. Clearly, much more work is required to interconnect devices and appliances within the home, and in turn, with the outside world (Figure 5.2). At the same time, the home environment provides a unique platform for research and innovation aiming to combine smart devices, sensors, personal/body area networks, and internet connectivity. The networked digital

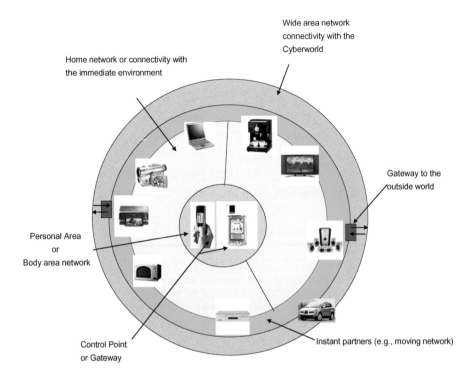

Fig. 5.2 Interconnected devices in the future home network.
Source: Dixit and Prasad [5].

home is an important link in the chain toward a portable user-centric digital ecosystem and should be considered alongside of parallel efforts in the field of personal and wide area networking.

5.1.2 Integration Challenges

Although there is much benefit to be gained by integrating the various communication, entertainment, home media and security systems in the home, there remain significant challenges. Integration eliminates duplication of control mechanisms and associated power consumption, interfaces and power supplies. It would also help limit untidy cabling. Still, most wireless applications in this field are fairly immature and in the "early adopter" or hobbyist phase of product diffusion. There is no "one-stop shop" that currently makes good use of system integrators for digitizing the home.

One of the main problems is that standardization of interfaces remains limited. Implementation would require incompatible technologies to work together. Indeed, most technologies in use in the home are promoted by specific industry alliances: some are widely accepted standards, such as WLAN (e.g. Wi-Fi), while others such as Z-Wave (Zensys) are proprietary specifications (Table 5.1). Some provide internet connectivity throughout the home and others simply replace wire or cabling over short distances (e.g. Bluetooth and infrared).

The different home wireless networks on the market today have at least one characteristic in common: they are connected to an external wide area

Table 5.1 Selected wireless technologies in use in the home.

Technology	Description
WLAN	Connection to a local area network (LAN) through a wireless (radio) connection, as an alternative to a wired local area network. The most popular standard for wireless LANs is the IEEE 802.11 series, such as Wi-Fi, or 802.11b. The term Wi-Fi is sometimes mistakenly used as a generic term for wireless LAN.
Bluetooth	Wireless protocol enabling the transmission of signals over short distances between mobile phones, computers, and other devices. Primarily used as replacement for wires and cables, e.g., wireless headsets. Also suitable for wireless PANs.
Infrared	Infrared data transmission is used for short-range communication between computer peripherals and personal digital assistants. Infrared devices typically conform to standards published by the Infrared Data Association (IrDA).
ZigBee	Open industry specification based on IEEE 802.15.4 standard. Although Zigbee supports slower data transmission rates than its competing specifications, it consumes significantly less power, and is particularly suitable for Wireless PANs.
DECT	DECT stands for Digital Enhanced Cordless Telecommunications and is an ETSI standard for digital portable phones (i.e. cordless home telephones). It can also be used for wireless broadband data transfers.
ONE-NET	Open-source standard for wireless networking, designed for low-cost, battery-operated, low-power control networks, for applications such as home automation, security, monitoring, device control, and sensor networks.
Insteon	Dual-band mesh technology developed by the US company Smart Labs, that employs power lines (AC) and a radio-frequency (RF) to enable home automation networking between devices and appliances that typically work independently to communicate with and automate home electronic devices and appliances, which normally work independently.
EnOcean	Single solution developed by German company EnOCean, for energy harvesting sensors, and RF (radio frequency) communication for building and home automation, lighting, industrial, automated meter reading, and environmental applications.
Z-Wave	Wireless communication protocol developed by Danish company Zensys, and backed by the Z-Wave Alliance. Designed for low-power and low-bandwidth appliances, such as home automation and sensor networks.

network, e.g., global mobile network (like 3G and GSM), fixed network (PSTN and ADSL). But although progress has been made on the networking of personal computers and audio–visual equipment, little has been done in the area of connecting white goods and appliances to the network. There is also little, if any, integration between appliances, entertainment consoles, and alarm systems.

Another challenge is the presence of competing solutions for the same application. For example, although digital TV can be taken "off-air" or delivered over cable, IPTV programming requires a TV set-top box for viewing and cannot be viewed on a home computer. In addition, Internet TV (i.e. streamed television programming over the internet) is becoming more and more popular as more programming goes on-line and greater bandwidth becomes available. Internet TV also allows broadcasting and video production companies to interact directly with viewers, thereby avoiding the involvement of network operators. This "over-the-top" business model is likely to better meet the needs of consumers, but there is a clear business conflict between network operators and emerging "over-the-top" service provision.

5.1.3 RFID, Sensors and the Home

Vital to the realization of the digital home is the standardization of the interfaces within it, from sensors to control units that can be connected to external networks such as mobile or internet networks.

Presently, sensors rarely provide information other than an indication of events (e.g., changes in environmental conditions or status) that they are designed to detect (see Section 3.4). In a networked environment, other useful information such as the specific location of the event can only be provided by integrating identification and communication capabilities in the sensor. RFID provides just this capability. Sensors that integrate RFID tags are now commercially available, though not widespread (e.g. the temperature sensor marketed by Identec) [6].

An RFID system can provide a unique platform for transmitting data between sensors and a central controller. Tags on sensors can communicate with RFID readers, which can then transmit data to the controller through wireless networking technologies like Zigbee, or those more widely available, like WLAN.

5.2   The AGE@HOME Platform: Objectives and Overview

5.2.1   Background

Analysts predict that significant innovation and take-up in the RFID market will occur in the healthcare sector in the near to medium term. However, although much of the literature points to the undoubted benefits of RFID for patient care and monitoring, the technology has yet to be widely adopted. This is due to a number of different factors, including, *inter alia*:

(a) lack of global standardization;
(b) market fragmentation (there is a variety of stand-alone solutions on the market);
(c) lack of awareness of the uses and benefits of RFID;
(d) lack of well-developed business models (cost and return on investment);
(e) lack of integration with existing electronic systems in hospitals and in the home;
(f) privacy concerns.

Given these concerns, it is proposed that RFID in the home should be further examined and applied to the independent living context. From a business perspective, the successful adoption of RFID in the home could help stimulate mass acceptance and deployment of a technology that has been plagued by user mistrust, privacy concerns, and market fragmentation. For the aged, RFID systems have the potential to provide independence, self-sufficiency, and enhanced quality of life. Moreover, and in sharp contrast to hospital or nursing home care, independent living solutions can lead to substantial savings for the increasingly penurious healthcare systems of many countries.

5.2.2   Objectives

The AGE@HOME platform is intended to monitor the activities of elderly persons and their homes to determine unusual activity/inactivity that may indicate illness, injury or incapacitation, as well as any unusual or unsafe environmental conditions arising in it. AGE@HOME is intended to be of greatest benefit for elderly single member households.

AGE@HOME is based upon automatic data capture and identification technologies, such as RFID and sensors, and these create a simple wireless system for independent elderly living. The system detects any unusual activity in a home and alerts caregivers or public health service providers of critical situations that might develop in the home. Existing systems and technologies are evaluated to better understand how they may be combined with more recent technology to achieve this important goal. A decision-support model to determine unusual activity in the home is presented as a pivotal element of AGE@HOME.

The designing of the AGE@HOME platform raises the following central questions:

(1) What kinds of automatic data capture systems are best suited to the homecare environment in order to detect unusual activity, or lack of activity?

(2) What kind of wireless network is best suited to the home environment, for low-bandwidth monitoring and data transmission purposes?

(3) How best to configure a decision-making support model for determining activity/inactivity in the home?

(4) How can data from RFID, traditional infrared sensors, and other types of sensors be combined effectively?

These can be raised in the larger context of how such a system might be viewed in the larger context of the so-called web 2.0, social networking tools and virtual media to enhance the independent living environment.

The motivation for designing such a platform is manifold, some of which has been described above. RFID has been classed in some circles, notably the press, as belonging to a growing group of "surveillance" technologies. Its use in the home, such as the AGE@HOME platform, may serve to assuage some of these concerns. Moreover, mass deployment of item-level tagging and sensors could benefit from use by potential early adopters, or lead users, such as the elderly. Ideally, platforms for AGE@HOME should be cost-effective, interoperable, and accurate. The objective is to create a system to empower caregivers and enable the elderly to be independent of nursing home care for as long as possible, through monitoring of home environments and resident activity. AGE@HOME may also generate ideas for spin-off uses of these technologies in the home.

Given its objectives, AGE@HOME enhances the potential for the application of existing and emerging technologies in line with crucial public policy initiatives. It fits squarely within European Union policies and research agenda on RFID, the Internet of Things, ageing and independent living for the elderly. Privacy concerns, which are increasingly public policy priorities, are also borne in mind through the possible use of a stand-alone system and the restriction of usage and data access to the elderly and their caregivers. Service providers that might manipulate complex levels of personal data will need to abide by data protection principles in the provision of services and marketing activities [7]. Public services are engaged only if required and in extreme situations.

The theoretical approach put forward here could be tested through various configurations in a living laboratory or alternative experimental context.

5.2.3 System Overview

The AGE@HOME system is composed of RFID readers, identification tags, different types of sensors, and is managed by an underlying decision-support system. It avoids the use of always-on surveillance cameras as well as the use of more costly robotic technologies, though these latter may be added in future systems. Environmental sensors are required to monitor water usage and temperature. Available devices, such as fire and flood detectors, are integrated into the total home monitoring system. Motion sensors are necessary for the detection of resident motion within the home. Pressure sensors are required for monitoring the use of furniture, such as the bed, or the sofa. A mechanism to determine the opening and closing of doors is required to indicate the potential consumption of food, e.g., the fridge. The system assumes that the resident wears a small bracelet containing an RFID tag and that any pets are tagged with RFID tags. Other items of importance, such as asthma inhaler or glucose meter, eyeglasses and mobile phones would also be tagged.

For ease of use and installation, data from the sensors is transmitted wirelessly. The AGE@HOME approach, as mentioned above, is to couple an RFID tag with each sensor, to enable simple communication between sensors and RFID readers that are strategically placed in each room. As sensor data has to be provided continuously to readers, active RFID tags will be required, for passive tags require the reader to trigger an event. Active tags will also facilitate the localization of important items (as discussed in Section 3.2.2) that have been tagged with RFID tags (e.g., mobile phones and keys), including medical and mobility aids (e.g., glucose reader, eyeglasses). The use of UHF RFID systems in the 433 MHz will ensure maximum read range.

RFID readers and wireless motion sensors are placed at the front door and at important points, e.g., the entrance of each room, and connected to a central decision-support system. A water mains sensor, a flood detector and fire detector(s) are also installed. Thermostats are present in the main living and sleeping areas. A pressure sensor is placed in or under the resident's bed mattress. A magnetic read switch or other type of sensor to determine the opening and closing of the fridge door is also positioned. Finally, a door lock sensor is placed on the front door.

Data from sensors is initially communicated to the RFID tag, which in turn sends its data to a reader in each room. A second wireless networking technology would enable the readers to communicate with the central controller (typically a personal computer). In order to enable connectivity between the readers and the controller, one option is to rewire the house with Ethernet. Alternatively, use might be made of power line communications. But each of these options is costly and relatively inconvenient. Zigbee is a more interesting alternative, given its use in hospital settings, but its use is limited and has yet to reach the mass market, with equipment available at high cost. As mentioned above, viable wireless networking technology that could connect readers with a central controller, and one which is already widely used and easy to install, is WLAN. WLAN-enabled readers are already in use in hospital settings (see Section 3.3). Figure 5.3 illustrates the AGE@HOME network configuration.

Equipment required for the AGE@HOME independent living platform is listed below:

- WLAN-enabled RFID readers.
- Active RFID tags.
- Wireless passive infrared (IR) sensors.
- Water meter sensor.
- Flood sensor.

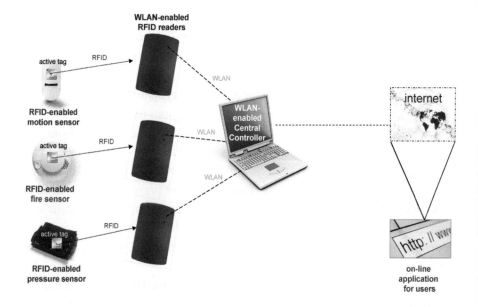

Fig. 5.3 AGE@HOME network configuration.

* Thermostats.
* Fire detector.
* Pressure sensor.
* Contact switches.
* Mobile phone.
* Central control box (including programmable logic controller).
* Gaming console (e.g. Nintendo Wii) with internet connection.

The decision support system receives signals from sensors, via the RFID readers, indicating that they have been activated. It analyses this information and generates alarms to care providers as appropriate using wireless or Internet access technologies. The system is designed so that intrusive video surveillance is not required to monitor household activity. The AGE@HOME decision-support model is further elaborated in Section 6, together with an informal graphical Specification and Description Language (SDL) description. An illustration of AGE@HOME in a typical one-bedroom apartment is laid out in Figure 5.4.

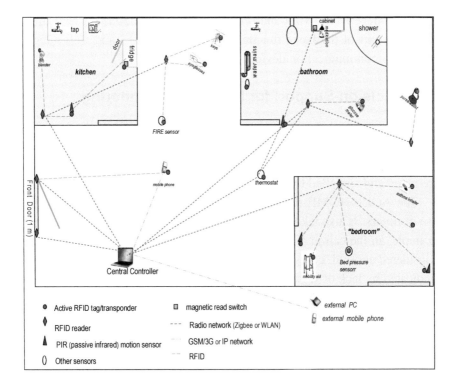

Fig. 5.4 Illustration of AGE@HOME in a typical one-bedroom apartment.

To cater to habits that are natural to the elderly, the use of a television set rather than a personal computer is recommended. The use of a device such as a "sensor bar" together with a corresponding remote control, similar to that provided with the Ninendo Wii console, may be a viable input mechanism. The Wii console and its remote (informally known as a Wiimote) uses a combination of Bluetooth, an accelerometer and infrared to detect motion. The console is WLAN-enabled for internet access. Whatever interface is used, the application should be available easily online through secure remote access. System users would then be able to navigate that web site easily through the use of a PC mouse, or by pointing devices at the television screen, not unlike the TV remote control. The advantage of using such a console is the possibility of including cognitive and lifestyle gaming for elderly system users. This would transform daily tasks into games, combining entertainment with health and lifestyle applications (see Section 6.2).

The independent living and security capabilities described here must be seen as a part of the basic functions of home networks and well integrated with other communication devices in the home.

5.3 Decision-Support for Behavioral Monitoring

Data about resident activity and environmental conditions of the home needs to be recorded on a continuous basis. This data is used to build a normal or reference model. Activity is compared to this model and significant deviations are noted and signalled by alarms.

The AGE@HOME model is based on sensor input combined with the use of timers. It also takes into account whether the resident is at home. For example, an indication that water is not being used while the resident is in is considered unusual. Likewise, a lack of physical activity for prolonged periods (indicated by lack of signals from motion detectors) in such circumstances is also considered abnormal.

The system is designed to alert caregivers, in the form of SMS, e-mail or voice message, depending upon situations and preference. For instance, alerts are sent out when the resident:

- is not at home in the morning (e.g. at 5 am);
- leaves the home;
- returns to the home;
- is out too long (e.g. more than 18 hours) ;
- has not left the house for a number of days (e.g. 10 days).

The resident wears an identifying RFID bracelet, which together with motion and door lock sensors, determine when the resident enters and leaves the home.

A pressure sensor is used to indicate when the resident is in bed and the caregiver can be alerted by SMS or e-mail if the resident has been in bed for more than 12 hours. If the resident continues to be in bed for 24 hours the emergency services or a doctor may be alerted in addition to the caregiver.

A sensor at the water mains detects water flow. Water flow is a good indication of activity in the home. Excessive use of water can indicate a problem with the home, or neglect on the part of the resident, and the caregiver is alerted if water has been running too long or if it has not been used for long periods while the resident is at home.

Motion detectors are placed in various rooms and in the bathroom. The caregiver can be alerted if there has been no motion for a prolonged period during the day, while the resident is at home. Emergency services or a doctor may additionally be alerted if the situation persists. Another procedure has been specified for detecting motion in the bathroom and uses an RFID reader to check if the resident is present. An alert is sent to the caregiver if the resident has been inactive in the bathroom for a period of time and a further alert is sent to the emergency services if inactivity persists.

Temperature sensors are used to check that the room temperature is within the range of 12–30°C. If the resident is at home and the temperature has been outside this range for more than a specified period, the caregiver will be informed. The resident and caregiver will also be alerted if the room temperature drops below zero. In the light of degrading environmental conditions and global warming, it is even more vital to ensure that the ambient temperature in housing for the elderly does not exceed tolerance limits.

A magnetic read switch is used to detect the opening and closing of the refrigerator door. The resident and caregiver are alerted if the fridge door has been left open too long and the caregiver is notified if the fridge has not been opened at all during a prolonged period while the resident is at home.

Alarms are raised if fire is detected and if flooding occurs. An intruder alarm is raised if motion is detected while the resident is out.

An additional capability of the AGE@HOME system is that users can request the system to indicate the location of an RFID-tagged object such as a pet or medical aid. This query can be effected over a simple user interface, and is particularly useful for those residents who often forget where they have left items of importance in the house.

6

AGE@HOME System Description, Implementation, and Implications

You know you're getting old when you stoop to tie your shoes,
and wonder what else you can do while you're down there.

— George Burns

This chapter provides a system description of the AGE@HOME platform and the results of the associated computer simulation. In conclusion, it discusses some of the challenges that arise for developers, users, and policy makers.

6.1 Specification and Description Language (SDL)

The use of SDL (Specification and Description Language) has been chosen for expressing the requirements and describing the functioning of the AGE@HOME system. This high-level object-oriented language used the world over was developed and launched by the International Telecommunication Union (ITU) in 1976 and is detailed in ITU-T Recommendation Z.100 and other Z-series recommendations. This Recommendation states, *inter alia*, that SDL:

(a) is easy to learn, use, and interpret;
(b) provides unambiguous specification for ordering, tendering, and design, while also allowing some issues to be left open;
(c) may be extended to cover new developments;
(d) is able to support several methodologies of system specification [1].

The universal popularity of SDL, especially for the specification and description of telecommunication systems, rests on its unambiguity and clarity. This language has been found particularly well-suited to applications such as:

- call and connection processing (for example, call handling, tele-phony signaling, and metering) in switching systems;
- maintenance and fault treatment (for example, alarms, automatic fault clearance, routine tests) in telecommunication systems;
- system control (for example, overload control, modification, and extension procedures);
- operation and maintenance functions;
- network management;
- data communication protocols; and
- telecommunication services.

Systems described using this language are often complex, event-driven, and real-time.

The language has two modes of expression — graphical (SDL/GR) and textual or phrase (SDL/PR). Either or both may be used to describe the same system. SDL is fairly general, such that even nontechnical persons can understand and use it. This naturally facilitates understanding between vastly different entities, such as clients, end-users, operators, designers, and manu-facturers. It originally focused on telecommunication systems. Currently, its areas of application include process control and real-time.

Although SDL is formally complete, nevertheless it can be used in an informal fashion at any level of description. Finalized documents cast in SDL are used as direct inputs for the generation of code, as is the case in the present research work. The graphic representation of SDL has been used in the section that follows to describe the AGE@HOME decision support system.

6.2 System Description

The AGE@HOME decision-support system is set out using graphical SDL language in more detail below. For simplicity, only a limited number of symbols are used. System behavior is represented in terms of states and transi-tions governed by the reception of input signals. Input signals can trigger other events such as sending output signals. There are three system states: Resident In, Resident Out, and Resident In/In Bed.

Timers calculate a number of durations, e.g., the length of time the resident is at home or in bed. Timers generate events when they expire or when they are canceled by specific input signals. The expiry of a timer in the system is an input signal in itself. Signals from sensors indicating that they have been activated become input signals. They may generate an output signal, such as an SMS transmission to a caregiver or the starting or stopping of a timer.

The system is based on collecting data from tags and sensors, and combining these with various timers. The system should be configurable so that users may set the duration of timers depending on individual habits (e.g., sleeping patterns) or specific situations (e.g., periods of absence from home). The central controller is programed to alert in many cases not only the caregiver but also the resident. This enables any cancelation by the resident of an alert, and thus the overall control of the system rests in the resident's hands. Emergency services are alerted only in extreme cases.

The figure below contains a list of symbols used in the SDL description of the AGE@HOME system. The description that follows examines each of its behavioral and environmental monitoring functions.

List of SDL (Specification and Description Language)
symbols used

	start
	state
	input
	output
	start timer
	cancel timer
	save
	decision

6.2.1   Residency Check

Through the RFID tag present in the bracelet, and RFID readers placed throughout the house, the system determines the resident's presence at home early each morning (e.g., 5 a.m.). If the resident is not detected, the query is re-sent within 10 minutes and then again in a further 20 minutes. This is to take account of any brief absences from the home (e.g., a walk or a trip to the bakery). If the resident is still not detected, an alert is sent to the caregiver accordingly (see Flowchart 1). The type of alert, SMS, email or voice message, can be set by the users themselves.

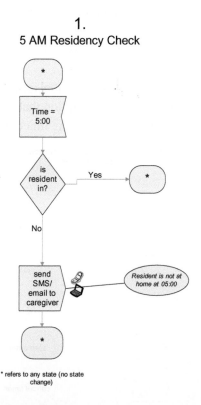

1.
5 AM Residency Check

* refers to any state (no state change)

6.2.2   Exit and Entrances Check

Upon the exit of the resident, an alert is sent to the caregiver, and the "resident out" timer is set at 18 hours. Whether an alert is sent each time the resident exits is a parameter that can be fed into the system upon installation. The exit of a resident brings to a stop other relevant timers, such as the "resident in" (RESIN) timer or timers related to motion, i.e., NOMOTION1 and NOMO-TION2 timers. The system state then changes from "resident in" to "resident out." This state will in turn govern the interpretation of data from home sensors and timers (see Flowchart 2A). The exit of a resident is signaled through the use of an RFID reader at the front door, coupled with data from motion sensors that may indicate a lack of motion throughout the home. The RFID bracelet itself may also contain motion sensors.

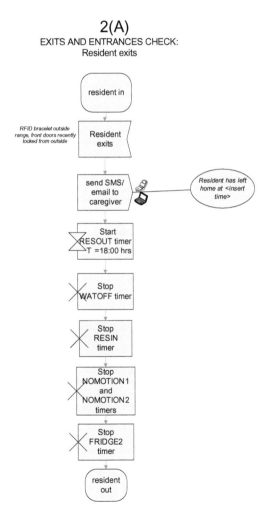

2(A)
EXITS AND ENTRANCES CHECK:
Resident exits

When the resident re-enters, an alert can also be sent to the caregiver to indicate the time at which the resident returned home (see Flowchart 2B). At that time, the "resident out" timer stops and a "resident in" (RESIN) timer starts and counts down from say 240 hours (10 days). The expiry of this timer will indicate that the resident has not been out of the house for the set period (see Flowchart 2C). This may be injurious to the resident's health. It may also be indicative of a possible decline of health. Therefore, an alert is sent to the caregiver upon expiry of the timer. The timer will stop anytime the resident exits the home, and timers can be configured for particular circumstances or for particular individuals.

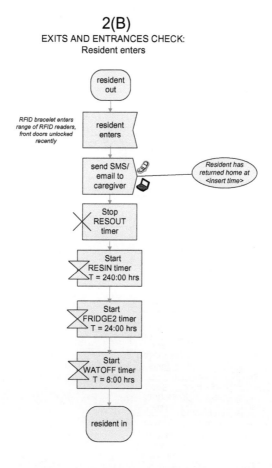

2(B)
EXITS AND ENTRANCES CHECK:
Resident enters

2(C)
EXITS AND ENTRANCES CHECK:
Resident in. Resident in too long.

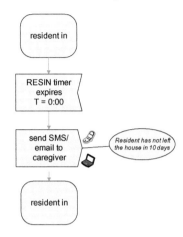

As mentioned above, while the resident is out, the 18 hour "resident out" (RESOUT) timer is running. Its expiry will indicate that the resident has been away from home for over 18 hours. An alert is then sent to the caregiver indicating the resident's time of departure (see Flowchart 2D).

2(D)
EXITS AND ENTRANCES CHECK:
Resident out. Resident out too long.

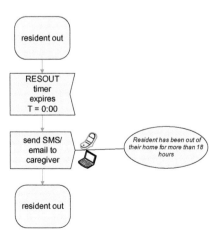

6.2.3   Bed Check

A pressure sensor (equipped with RFID tag) detects the presence of a resident in the bed. The sensor can be configured to reflect the resident's weight. However, a good default value should be 40 kg. This is in order to avoid activating the sensor when a weighty object (e.g., suitcase) is placed on the bed. When the minimum set pressure is detected in the bed, two timers are initiated (see Flowchart 3A): INBED1 (12 hours) and INBED2 (24 hours). Both timers stop when the resident is detected to be out of bed (see Flowchart 3B). Two timers are required for two corresponding alert levels. Where a resident has been in bed for 12 continuous hours, an initial alert is sent out to the caregiver and to the resident. A persistence of the situation for a further 12 hours will result in the alerting of emergency services, as this should indicate a serious medical problem (see Flowchart 3C). A second alert is also sent to the caregiver.

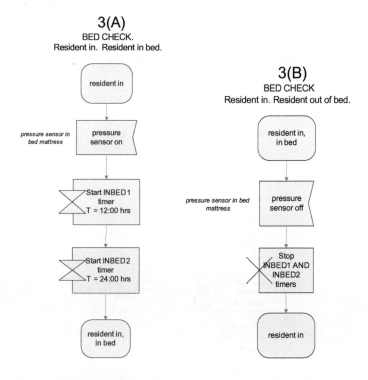

3(A)
BED CHECK.
Resident in. Resident in bed.

3(B)
BED CHECK
Resident in. Resident out of bed.

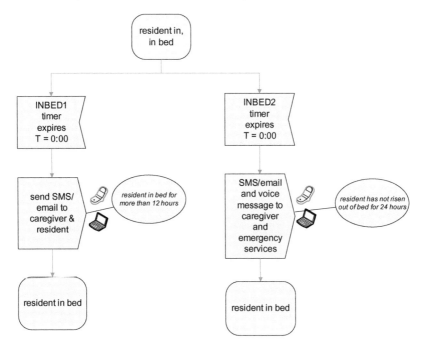

3(C)
BED CHECK
Resident in. Resident in bed too long.
(12 hours and then a further 24 hours)

6.2.4 Water Flow Monitoring

Use of the piped water system in the household is an important indicator of resident activity. When water turns on at the mains, the water meter sensor (equipped with RFID tag) is activated, starting the WATON timer, for 2 hours (see Flowcharts 4A and 4C). When the water turns off, the WATON timer stops and the WATOFF timer sequence is initiated, counting down from 8 hours (see Flowchart 4B).

The water sensor can determine whether there is continuous (unusual) water use, whether the resident is in or out (see Flowchart 4D). If the water has been turned on but not turned off for 2 hours, corresponding alerts are sent to both the caregiver and the resident. This alert is repeated for every 2 hours until water is turned on.

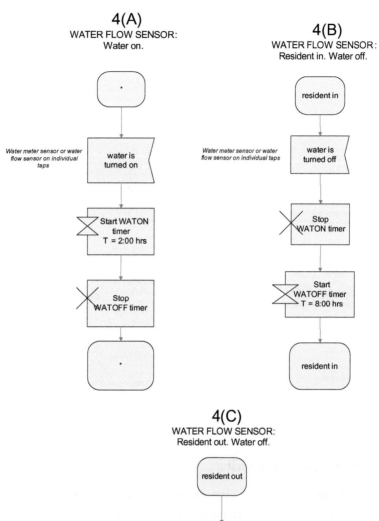

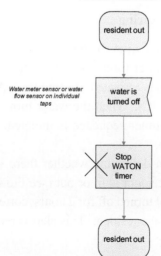

4(D)
WATER FLOW SENSOR
Water turns on and then does not turn off within 2:00 hrs

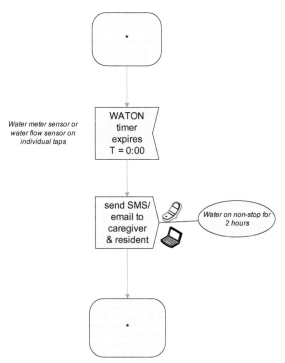

When the resident is in the home, the frequency of water use is considered an important determinant of the normalcy and health of a resident. If the WATOFF timer expires, the system checks the time, in order to take into account sleeping hours. If it is between 13:00 and 23:00, the WATOFF timer is re-set for 8 hours. If it is not during that time period, an alert is sent to the caregiver and the resident (see Flowchart 4E). Use of sensors on individual taps, for the purpose of differentiation, is a matter for future consideration.

4E.
WATER FLOW SENSOR
Resident in. Water not turned on for a continuous period of 8 hours

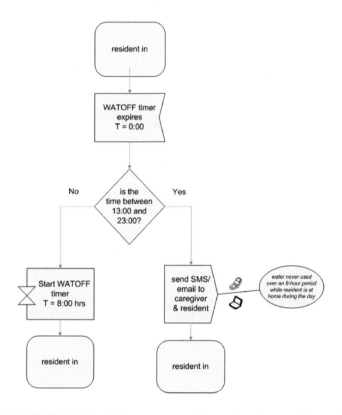

6.2.5 Motion Detection

General motion detection

Motion sensors are located throughout the house and monitor movement. Lack of movement is determined through the interpretation of timers. When motion is detected in any of the rooms in the house, two timer sequences are initiated: NOMOTION1 counts down from 5 hours and NOMOTION2 counts down from 12 hours (see Flowchart 5A).

5(A)
MOTION DETECTION
Resident in. Motion detected.

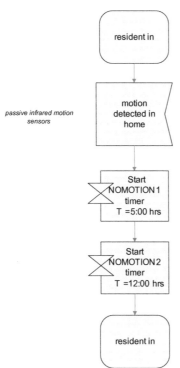

passive infrared motion sensors

resident in

motion detected in home

Start NOMOTION 1 timer T =5:00 hrs

Start NOMOTION 2 timer T =12:00 hrs

resident in

If no motion is detected by the motion sensors for 5 hours, the system checks if the resident is in bed. If the resident is detected in bed, no alert is required, and the "inbed" series of timers are in operation. If the opposite is the case, an alert is sent out to the caregiver and the resident. If the situation persists for a total of 12 hours (7 further hours), emergency services are alerted, and the caregiver receives a second alert (see Flowchart 5B).

When the resident is out, the motion sensors function as an intruder alarm (see Flowchart 5C). The return of the resident to the home is signaled via their RFID bracelet. Like most intruder alarm systems, authorized visitors should be able to enter the house in the resident's absence.

5(B)

MOTION DETECTION
Resident in. No motion detected for a continuous period of 5 hours
(and a further 12 hours)

resident in

NOMOTION	NOMOTION2
1 timer	timer
expires	expires
T = 0:00	T = 0:00

is INBED1 timer *T > 0.00 ?*   — Yes → resident in

No

SMS/email and voice message to caregiver, GP or emergency services

no motion for a continuous period of 12 hours while resident is in

send SMS/ email to caregiver & resident

no motion for a continuous period of 5 hours while resident is in

resident in

resident in

Detection of motion in the bathroom

Motion detected in the bathroom results in the initiation of two timers: BATH1 timer counts down from 3 hours and BATH2 timer counts down from 12 hours (See Flowchart 5D). If the first timer expires, a second check is conducted to verify via the RFID readers whether the resident is still in the bathroom. If this is the case, an alert is sent to the caregiver and the resident, as 3 continuous hours in the bathroom might indicate a physical problem. Upon expiry of the second timer (12 hours), another RFID check is conducted. If the resident continues to be detected in the bathroom, their caregiver receives a second alert, and emergency services are appropriately signaled (see Flowchart 5E).

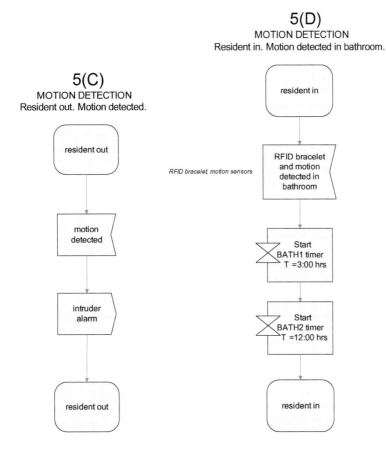

6.2.6 **Temperature Monitoring**

The home is equipped with a temperature sensor that detects temperature levels. When the resident is in the home, and the temperature readings fall outside a set normal range (e.g., 12°–30°C), a TEMP1 timer starts (see Flowcharts 6A and 6B). Where the temperature does not return to normal within 12 hours, an alert is sent out to the resident and their caregiver (see Flowchart 6C). The speed of the timer (i.e., its duration) is a function

5(E)
MOTION DETECTION
Resident in. Resident in bathroom for a continuous period.

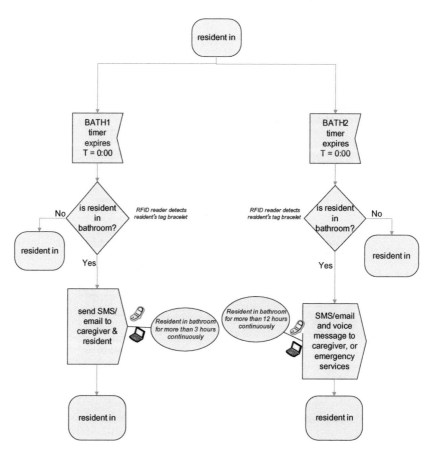

of the number of degrees above the preset maximum: if the temperature is 31°C, the timer will run at normal speed, but at 40°C or 45°C, the speed will increase so as to provide earlier alerts. When the resident is out, abnormally high temperature levels are not as much of a concern. But with temperatures below 0°C, both caregiver and resident should be alerted (to indicate risk of freezing) if the temperature levels persist for over 1 hour (TEMP2 timer) (see Flowcharts 6D and 6F). Upon return of temperature to levels above freezing, the corresponding timer stops (see Flowchart 6E).

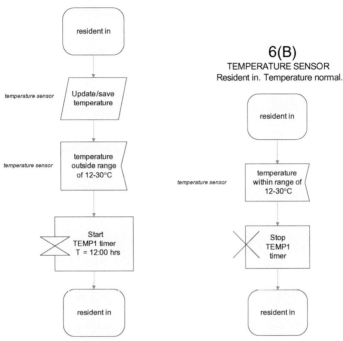

6(A)
TEMPERATURE SENSOR
Resident in. Temperature abnormal.

resident in

temperature sensor — Update/save temperature

temperature sensor — temperature outside range of 12-30°C

Start TEMP1 timer T = 12:00 hrs

resident in

6(B)
TEMPERATURE SENSOR
Resident in. Temperature normal.

resident in

temperature sensor — temperature within range of 12-30°C

Stop TEMP1 timer

resident in

6(C)
TEMPERATURE SENSOR
Resident in. Temperature abnormal for over 12 hours.

resident in

TEMP1 timer expires T = 0:00

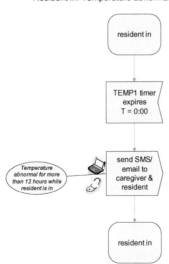

Temperature abnormal for more than 12 hours while resident is in

send SMS/ email to caregiver & resident

resident in

6(D)
TEMPERATURE SENSOR
Temperature highly abnormal.

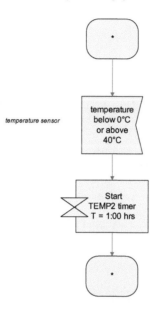

6(E)
TEMPERATURE SENSOR
Temperature normal.

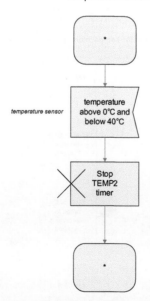

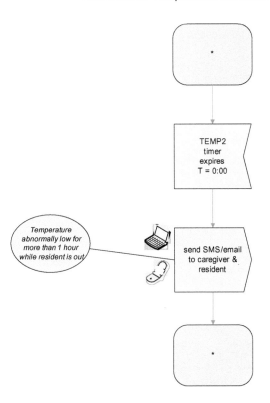

6(F)
TEMPERATURE SENSOR
Resident out. Temperature abnormal for over 1 hour.

6.2.7 Fridge Door Monitoring

Sensors in the fridge door monitor activity or lack of it. In association with timers, the system determines regularity of fridge use. If a fridge door has been left open exceeding a specified period, say 6 hours (FRIDGE1 timer), an alert is sent out to the caregiver and the resident (see Flowcharts 7A and 7C). Similarly, if the fridge has not been opened for a specified period, say 24 hours (FRIDGE2 timer), an alert is also sent out to the resident and their caregiver (see Flowcharts 7B and 7D).

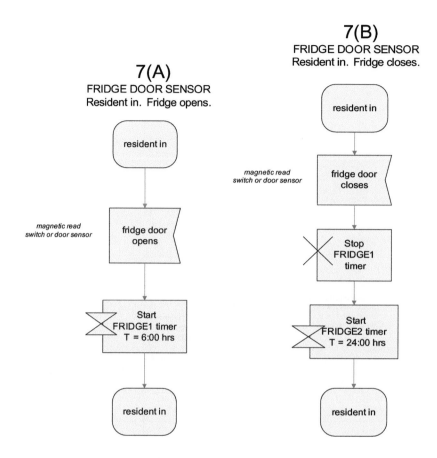

7(A)
FRIDGE DOOR SENSOR
Resident in. Fridge opens.

7(B)
FRIDGE DOOR SENSOR
Resident in. Fridge closes.

7(C)
FRIDGE DOOR SENSOR
Resident in. Fridge opens and stays open for 6 hours.

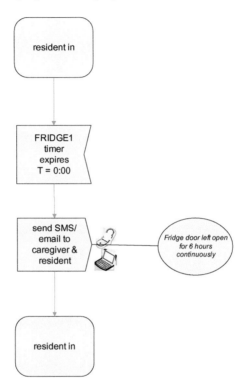

7(D)
FRIDGE DOOR SENSOR
Resident in. Fridge is not opened in 24 hours.

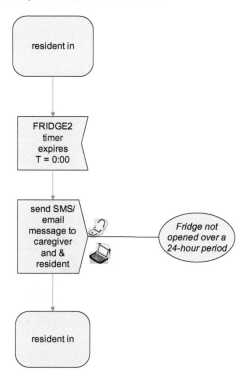

6.2.8   Fire Sensing

Fire sensors in the home detect the presence of fire, with alerts sent out in the event of fire to the resident, their caregiver and the fire department (see Flowchart 8).

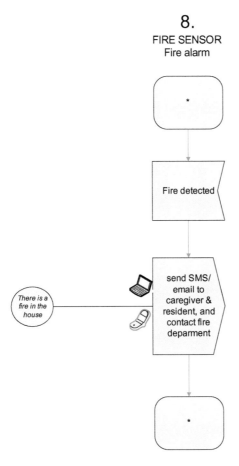

8.
FIRE SENSOR
Fire alarm

* refers to all states (no state change)

6.2.9   Flood Sensing

A flood sensor in the home detects the presence of a flood, with corresponding alerts sent out to the caregiver and the resident (see Flowchart 9).

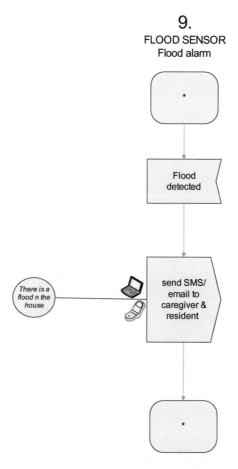

9.

FLOOD SENSOR

Flood alarm

*

Flood
detected

There is a
flood n the
house

send SMS/
email to
caregiver &
resident

*

refers to all states (no state change)

6.2.10   RFID Localization of Resident and Objects

RFID readers can detect the presence of the resident, and of tagged objects, that come within their read range. This information can be saved in the central controller (see Flowchart 10A) for retrieval when required (see Flowchart 10B). The different signal strengths of the various readers are compared to determine the closest reader. Through the user interface, the object can then be displayed on a map for the user to view the most likely location of the misplaced object.

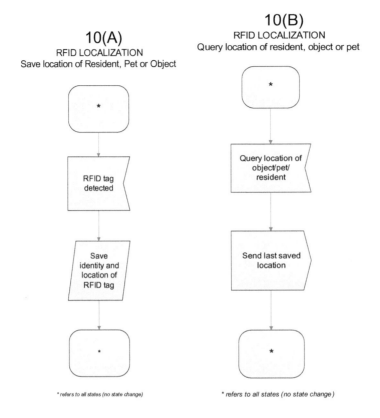

10(A)
RFID LOCALIZATION
Save location of Resident, Pet or Object

10(B)
RFID LOCALIZATION
Query location of resident, object or pet

* refers to all states (no state change)

* refers to all states (no state change)

6.3 Software Implementation

This section describes the simulation and control software for the AGE@HOME independent living platform. The software was developed using MATLAB programming language and incorporates the decision support system as described in SDL in Section 6.2. It includes sensor and timer controls, as well as a simulation platform. It was designed to demonstrate the operation of the AGE@HOME independent living platform, and works as a finite state machine operating in cycles. In each cycle, it reads the latest sensor data, updates the timers, and sends out any corresponding reports. The system administrator can engage with the simulation and data from the sensors through graphical user interfaces. In so doing, sensors can be activated or deactivated, and current timer values verified, thereby imitating a real-life scenario.

The software consists of several components, each with their own respective functions, namely:

- decision logic and timer control,
- control interfaces,
- object localization,
- simulation.

Each of these is discussed in more detail in the following section.

6.3.1   Decision Logic and Timer Control

The AGE@HOME system requires input from sensors. It uses this input to feed into the decision logic described in the SDL diagrams in Section 6.2. This enables it to control and update timers, and to send out alerts when certain timers expire. Reflecting this logic, the software, as mentioned above, is a finite state machine operating in cycles. This machine is controlled by a master timer which regulates how often the software reads new sensory input.

Sensor check
During each cycle, the machine checks for new sensory input (either directly listening to input ports or by reading a log file containing all recent incoming messages). It uses this input to change the states assigned to each sensor: "active" or "inactive." States switch from one state to the other immediately after a message of activity is received from the sensor. However, since sensors do not send any data while they are awaiting new input, a time gap of inactivity has to pass before the sensor's state switches from "active" to "inactive." Within the software, and the documentation related to it, this initial task of checking for sensor activity and correcting for sensor states is referred to as the "sensor check." The software performs these checks and then updates the sensory input of the relevant timers in operation, i.e., it informs each timer whether their sensor is currently active or inactive.

Timer update
The second task performed during each cycle is the update of timers. To accomplish this, the software checks the status of all existing timers. It verifies the conditions of the sensors to which the timers are associated and checks all the relevant states (i.e., resident's location, whether the resident is in bed,

Table 6.1 Timer object structure.

Variable	Value
use	1
time_limit	43200 [seconds]
states	$\begin{bmatrix} 1 & 0 & 0 \\ 0 & 0 & 0 \\ 0 & 1 & 1 \end{bmatrix}$
action	[1 1 0 0]
activity	1
gap	60 [seconds]
bed	1
bathroom	0
name	"Resident in bed for 12 hours"
running	0
start_time	[ ]
current_time	0 [seconds]
last_input	[2008 08 17 11 52 36.04]
sending_report	[0 0]
displ	197.0170

etc.). It then updates the timers, stops those that meet certain conditions, and sends reports, if necessary, when any timer expires.

Timer objects used by the software are structures consisting of several fields, each containing some relevant data for the operation of the timer and for behavior monitoring (see Table 6.1). The "use" flag indicates whether the timer is currently in use and if it should be verified. The "time limit" field contains information on the period it takes for the timer to expire once it is triggered (provided the conditions that caused the trigger remain in effect). The field labeled "states" contains information on the conditions for timer advancement, pause or reset. Rows in the "states" matrix correspond to the following timer states: "running," "pause," and "stop." Columns in the "states" matrix refer to the following resident states: "ResOut," "ResIn," "ResIn and in bed."

The "action" field specifies which reports are to be sent in case the timer expires, i.e., this refers to the first two reports for the timer given in the matrix (value 1). The flag labeled "activity" determines if the timer is running while the sensor is active (value 1) or while the sensor is inactive (value 0). The "gap" field is given in seconds and indicates how large a gap in time is permitted between consecutive sensor reports, in order for them to be considered continuous. "Bed" is a flag indicating whether the timer is related to the bed sensor, and the "bathroom" flag indicates whether the timer is connected to the "resident in bathroom" state. "Name" refers to the string variable shown in the timer and history fields. Of course, the names that are chosen by the system administrator should be easy to understand and unique.

The field labeled "running" shows the current state of the timer: 0 if stopped, 1 if running, 2 if paused, 50–52 while sending reports. The "start time" refers to the point in time at which the timer started running. This is useful by way of reference: the display format is "year:month:day:hour:minute: seconds." This field remains empty if the timer is not running — it only generates a display if timers are running or paused. The "current time" variable shows how many seconds have elapsed since the timer started running. The "last input" refers to the time at which the related sensor was last observed in the active state.

The "sending report" field is active only while reports are being sent out during the timer expiry procedure, and controls the manner in which this information is displayed in the interfaces. The last value "displ" is a pointer to the memory location of the associated timer display box.

During the timer update, each timer is checked for its running conditions. State and activity tables are compared to the current state of the resident and the state of the sensors: the timer's state is then adjusted accordingly. If the timer began in a running state, and the new state requires that the timer continue to run, then its time is advanced by one cycle. For implementation purposes, the duration of one cycle is exactly one second.

In performing these two tasks — the sensor check and the timer update — the software undergoes the decision-making process described in the SDL diagrams set out in Section 6.2. The timers start, stop or reset, and appropriate reports are sent out, according to that decision logic. The software goes a step further, of course, as it enables a system administrator to adjust the entire decision logic for each sensor, notably the conditions for timer activation and duration, and for the sending of reports. The default values in the software are those described in the SDL diagrams, but it is important to note that different combinations can be configured. This is achieved through the software's main configuration menu. The system can thus be designed to better fit a variety of environments and specific end-user needs. It also gives the system administrator the freedom to pause the timers if certain conditions are met, or create very specific running conditions. The only fixed conditions are the special indicators for the "Resident in/out/resting" states, as these combine inputs from several sensors and would require a much more complex interface to implement.

6.3.2 Control Interfaces

The main simulation master board for the AGE@HOME platform is shown in Figure 6.1. It contains the main sensor controls, provides feedback on the state of sensors, and displays all timers used for the decision logic and monitoring.

For ease of use, the simulation board is divided into several sections or panels. The top most panel in the first (left) column is the "system power" panel which contains a timer box showing the current system time, a system power on/off button, and a radio or option button display that is used only to show the current state of the system. The system administrator is not able to interact with these elements.

Below the system power panel, the left column contains the controls and timers for all possible sensors used for the monitoring of the home. The suitably labeled pushbuttons can change the current state of the sensor. There are only

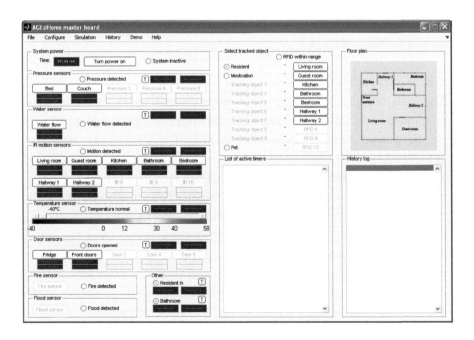

Fig. 6.1 AGE@HOME master board.

three possible states: "active" (which means that the sensor is sending periodic reports), "not active" (no reports are being sent), or "not in use." If the sensor is not in use, this is because it has not been physically deployed or because the system power is turned off. In such a case, the sensor control and its timers will appear grayed out and cannot be manipulated by the system administrator.

Timer display

Timers are displayed as dark gray boxes with the timer duration written inside in the following format: "hours:minutes:seconds." Timers have four possible states: "inactive," "running," "pause," and "reporting." Each of these states is color coded. If the numbers in the timer display appear in dark green, the timer is not running and is in the "inactive" state: it is simply waiting for the sensory input that will trigger it. Numbers appear in a brighter green to indicate that the timer is running and counting down the time. Numbers appear in yellow to denote a timer in the "pause" state: this means that the timer has been triggered, but that the current conditions have caused it to temporarily suspend its countdown: for example, the "no movement" timer is paused while the resident is resting on the bed. Numbers appearing in red (dark and bright red) denote that the timer has run out and is now sending the appropriate reports for that event.

Sensor display

The "pressure sensors" panel controls the pressure sensors. These sensors, as mentioned in earlier sections, react to a certain applied force, e.g., a bed sensor (see Section 6.2.3 and associated SDL diagrams). In a deployed system, if one of the pressure sensors detects that it is in use, its RFID tag will send a periodic message to the nearest RFID reader, which will then forward this information to the main controlling computer. The computer either feeds this information directly to the software, or writes it down inside a log file which is periodically read by the controlling software during sensor check cycles. Once the software realizes that the sensor is active, it will indicate this by changing the sensor button color to yellow, and updating the controller's database with the latest information. In simulation mode, the system administrator can control whether a specific sensor is sending active reports by clicking on the button: this action results in the state switching from "active" to "not active" or vice-versa. In demonstration mode, the simulation controls the states of the sensors according to a set of behavior rules and probability functions. Several sensors can be active at the same time.

The radio or option button labeled "pressure detected" indicates whether any of the pressure sensors are to be considered active at any given time. If they are active, the radio button will appear in the "active" state: the circle adjacent to it will be filled and the text background will become green. These particular radio buttons cannot be manipulated by the system administrator. The two timers that appear in the top right portion of the panel are the main timers. They display the time data on the pressure timer that is most likely to run out first. The "T" button to the left can be used to configure the timers to work with the "pressure detected" radio button instead, activating or deactivating it in relation to pressure detection over the whole sensor group (i.e., in the case of multiple pressure sensors).

Other sensor panels follow similar rules and display information in the same manner. The only exception is the temperature sensor panel used for monitoring ambient temperature as described in Section 6.2.6. The temperature sensor is the only type of sensor that is able to send data that goes beyond a mere periodical activity report. The measured ambient temperature of the home is embedded into its tag address: as the temperature changes, changes are made to the last four hexadecimal numbers in the tag address. In the master board, the temperature sensor panel thus displays the latest measured temperature, both as a number in the top left part of the panel, and as a position in the horizontal slider. During simulation mode, this slider can be manipulated by the system administrator to indicate a change of the ambient temperature in the house. The colored bar under the slider has six labels, showing the boundaries used to change the states of temperature reporting. The leftmost and rightmost numbers show expected minimum and maximum temperature. The pair closer to the center (second and fifth label) indicate the border between extreme temperatures (i.e., lower than the second or higher than the fifth number) and abnormal temperatures. The innermost two labels (third and fourth) mark the temperature zone that is to be considered normal. If the temperature ventures into the abnormal or extreme zones, the temperature panel radio button changes accordingly. The temperature display is also color coded to show the currently accumulated heat in the environment, i.e., excess heat or the absence of heat building up over time, as compared to normal conditions.

Fire and flood sensors do not have timers, as they send reports to the resident, caregiver, and all the relevant intervention units right away (as described in Sections 6.2.8 and 6.2.9).

The panel labeled "other" displays the resident in and resident out timers, as well as the state of the special bathroom timer. Both of these use combined sensor input to determine the resident's current location and condition, for their decision-making process. They are governed by logic described in the SDL diagrams in Sections 6.2.1, 6.2.2, and 6.2.5.

Tracking household objects

The second (middle) column of the main configuration board contains the panel for tracking objects within the home. The radio buttons for selecting the tracked objects are placed on the left side of the panel. The RFID tags that have not been deployed are grayed out. Clicking on any of the available radio buttons automatically selects the associated RFID tag. The right side of the object tracking panel includes pushbuttons for the RFID readers: each showing the reader's name. If any readers are not in use (or if the system is in the "off" state), those readers are grayed out and do not allow user/administrator interaction. If any of the readers are displayed as active, these readers are able to detect the signal of the currently selected object. If information regarding the received signal strength is available, it is shown to the left of each reader's button and the readers are color coded from the strongest (green) to the weakest (red), with combinations of yellow in between. The radio button labeled "RFID within range" acts as a verification and indicates if any of the RFID readers are able to detect the currently selected object. This particular radio button is not under the control of the administrator. Section 6.3.3 describes in more detail the object localization function of the software.

Timer list, map panel, and history log

The lower part of the second column contains a comprehensive list of all active timers (all that are currently running). It displays the names of the timers and the time at which they are set to expire.

The third (right) column contains the map panel, which displays a floor plan image of the monitored environment. This is where the information on the probable location for a particular tracked object is shown, along with the position of readers that are able to detect the tag on a particular tracked object.

Below the map appears the history log list. This is where all reports and actions taken by the software are recorded, from sent SMS messages to the resident and caregiver to fire department and intruder alarm data.

Configuring sensors and timers

The "file" menu on the main configuration board gives access to open and save functions that can be used to set up the environment rapidly and efficiently. This menu can also be used to recall and carry on previous simulations, or recall the accumulated data in the monitored environment. The "configure" menu provides a system administrator with the ability to fully customize the system, including all sensors and timers, in order to make it most suitable to a particular environment or to special circumstances. The "simulation" menu comprises the simulation parameters for choosing the mode of operation: testing, user-controlled, demonstration, or actual/real sensor input. The "history" menu presents the monitored data and allows it to extract the behavioral patterns after processing. The menu labeled "demo" deals only with the parameters for the demo simulation of the environment, in which all controls and sensors behave according to set probability functions. The "help" menu opens up a document describing the use of the software and the functions of each control and display.

Through the sensor configuration menu, the system administrator is given the option to change information on RFID readers, and the pressure, motion, water, door, temperature, flood and fire sensors and associated timers. Selecting an option opens up a user interface which contains the list of all sensors of that type, with their respective data and timers. An example of this, the RFID reader configuration board, is given in Figure 6.2. A total of 10 RFID readers can be set up inside the home environment, and each reader's data is represented by a row in the configuration window. The first column is the RFID reader name. This name can be edited, in order to enable users and administrators to better differentiate between readers: a name such as "kitchen" is easier to understand than tag addresses used for recognition and communication between sensors and readers. The second column represents the tag address of the sensor and is used to sort incoming reports. All reports which carry that particular tag address are considered to be reports from the associated sensor. It is possible to assign the same address to several sensors, and the software would then consider them to be the same sensor. This gives the administrator

RFID_configuration_board

RFID reader configuration

	RFID reader name	RFID reader adress	RFID reader positions		Reader in use	
RFID 1	Living room	RFID000001	(4.5m , 13.6m)	Placement	☑ In use	T
RFID 2	Guest room	RFID000002	(15.8m , 14.2m)	Placement	☑ In use	T
RFID 3	Kitchen	RFID000003	(4.4m , 5.6m)	Placement	☑ In use	T
RFID 4	Bathroom	RFID000004	(12.5m , 7.4m)	Placement	☑ In use	T
RFID 5	Bedroom	RFID000005	(14.1m , 4.1m)	Placement	☑ In use	T
RFID 6	Hallway 1	RFID000006	(8.4m , 3.9m)	Placement	☑ In use	T
RFID 7	Hallway 2	RFID000007	(15.3m , 10.7m)	Placement	☑ In use	T
RFID 8	RFID 8	RFID000008	(- , -)	Placement	☐ Inactive	T
RFID 9	RFID 9	RFID000009	(- , -)	Placement	☐ Inactive	T
RFID 10	RFID 10	RFID000010	(- , -)	Placement	☐ Inactive	T

Status message

Done

Fig. 6.2 RFID reader configuration board.
Note: The system administrator can change the name, address, placement, usage, and timer parameters for each available reader.

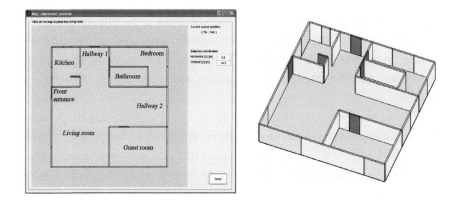

Fig. 6.3 Map window panel.

the possibility to trigger and stop more than just two timers with one sensor.

The third column labeled "RFID reader positions" displays the positions of any active readers in use. Clicking the button labeled "placement" opens a user interface displaying a map. Figure 6.3 (left) shows this two-dimensional representational map of the home environment. Figure 6.3 (right) is a 3D image of that same environment. Using this interface, the system administrator can select a point on the map by pointing and clicking with the mouse and then pressing "done." That position is then saved as the selected RFID reader's location. This information is essential for object tracking and localization. It has to be entered for each sensor in use before the system administrator is able to close the RFID configuration board and return to the main interface.

The fifth column in the RFID reader configuration board (as displayed in Figure 6.2) contains checkboxes that indicate which sensors are currently in use and which sensors are to be left out during the update phase (i.e., those that are inactive). The last column contains buttons labeled "T," each of which is linked with the RFID reader in its corresponding row. Clicking these buttons opens up the timer configuration window, allowing the system administrator to change all parameters for the associated timers.

The temperature sensor is a special case in that it sends reports only for every 30 seconds regardless of the value of the measured temperature. It also has a changing tag address, the last numbers of which carry the information on the value of the measured temperature. The timers associated with this sensor are triggered when the temperature exceed the limits set for abnormal or extreme temperatures. Therefore, the temperature configuration board uses an interface with two editable fields, to reflect the maximum range of temperature measurable by the sensor (Figure 6.4). It also includes four controls allowing the system administrator to establish triggers for temperature timers within

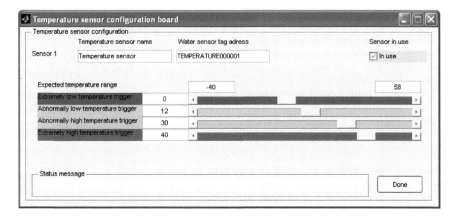

Fig. 6.4 Temperature sensor configuration board.
Note: The system administrator can define the abnormal and extreme temperature limits, as well as the name and tag address of the sensor.

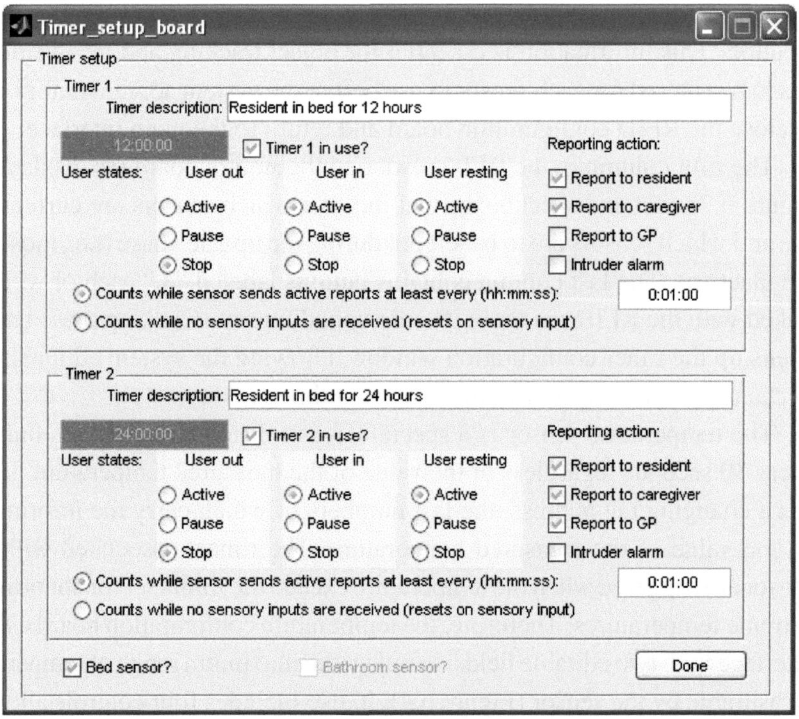

Fig. 6.5 Timer setup board for sensors and RFID readers.
Note: Timers can be turned on or off, and all of their parameters can be changed to fit a variety of needs.

this range by simply moving the sliders to the left (for a lower temperature) or right (for a higher temperature).

Timer objects are updated through the user interface shown in Figure 6.5. The timer setup board is brought up during the configuration of sensors in cases when the system administrator wishes to change timer parameters. Each timer setup board contains two timer objects that can be edited. This allows the system administrator to set one to run while the sensor is active, and the other to run while the sensor is inactive. It is also possible to assign both to the same activity cycle, but with different durations and different reports, e.g., one that generates a warning report to the caregiver, and then if the condition persists, a second more serious report to emergency services. This is considered sufficient for practically all cases. However, the system administrator can also create another object with two new timers assigned to it, and associate it with the same address. In this manner, more than two timers can be updated with data generated by the same sensor. This makes additional levels of reporting and alerts also possible.

Editing the "timer description" fields changes the timer name shown in the running timer list and history log. Below these fields, timer boxes for each of the timers are displayed. Unlike the ones in the main board, the timer boxes here are editable and allow the system administrator to enter the desired timer run time in a "hours:minutes:seconds" format, which is then converted to seconds and saved in the "time limit" field. The checkboxes next to them enable or disable the use of the timer. The radio buttons in the middle of each timer's configuration panel connect the timer actions to the resident's state. In other words, these are directly and graphically related to the matrix shown in Table 6.1.

By checking or un-checking the boxes marked under "reporting action," the system administrator can choose which reports are sent in the event that the timer runs out. These are saved in the "action" field of timer objects. The two radio buttons that appear lower down in each panel are mutually exclusive and determine the timer "activity" value (1 if running while sensor active, 0 if running while sensor inactive). The white editable field to the right of these two radio buttons represents the time gap used in the decision-making process: this is entered in the "hours:minutes:seconds" format. The bottom two checkboxes are used to associate the timers with either the "Resident in and in bed" state, or the "Resident in bathroom" state.

From SDL decision logic to software implementation
Figure 6.6 illustrates the relationship between the SDL logic diagrams set out in Section 6.2 and the software implementation. In particular, it shows how the control logic relating to the bed sensor from the SDL diagrams is implemented using the simulation software.

The state of the first pressure sensor is set to "in use," which means that it will be updated during simulation cycles, and that two timers have been assigned to it. The name of the sensor provides an adequate description for the system administrator to navigate more easily through all the controls, but it does not appear in any message or have any impact on the software execution. The tag address field should correspond to the tag address of the sensor, in order for the data to be received and analyzed by the sensor check routine.

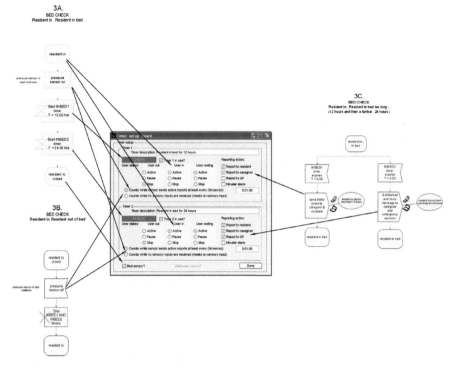

Fig. 6.6 Graphical illustration of how the SDL logic was implemented in the software: Timer Setup Board.

The placement window is used to position the sensor on the map (by clicking on the map spot where the bed is located). This creates a much more simple search mechanism, and produces a more detailed report on the behavior of the resident.

The timer setup board is the interface that the system administrator uses to enter all the data from the sensors. Since there are two timers in the SDL diagram, both timers have been enabled in the timer setup board. Their time limits have been set to 12 and 24 hours respectively, and their states reflect the starting states and queries in the SDL diagrams. The administrator only needs to read which reports need to be sent when the timer runs out and check the appropriate timer boxes. Both of these timers calculate the time the resident has spent in bed, so they both continue to run as long as the reports from the sensor continue to be received at intervals smaller than the associated time gap.

6.3.3 Object Localization

The software designed for the AGE@HOME platform enables residents to locate within the home any object that is equipped with an active RFID tag (see Section 6.2.10). The tags intended for tracking transmit a beacon signal every 2 seconds. This beacon signal is detected by nearby RFID readers and then forwarded to the central controller (PC). When a localization request is received, the software checks the most recent data received from the RFID tag in question, i.e., its last known transmission time and all other transmissions which occurred within a small time gap prior to the last transmission. The software is able to calculate the position of the wanted object within a certain area, by analyzing the following: the positions of the RFID readers, the strength of their beacon signals, and pre-calculated signal coverage maps (Figure 6.7).

During set-up, the system administrator initiates the signal coverage map calculation by placing the RFID readers on the map and saving the project. This procedure creates coverage maps for each of the localized RFID readers in use, and stores them for future purposes. This reduces all subsequent localization requests to simple matrix searches, which can be performed almost instantly. In this manner, the software maintains the precision of ray-tracing tools of the highest accuracy, while the response time is as short as it would be in simple and less accurate tools. Localization requests can therefore be processed within a fraction of a second, regardless of the complexity of the environment.

Using the measured signal strengths of the RFID readers and the coverage maps described above, the software is able to create probability maps such

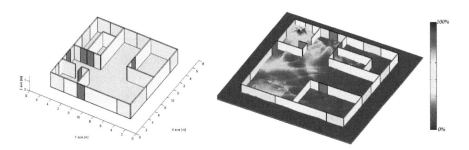

Fig. 6.7 Sample AGE@HOME environment with and without overlay of signal coverage map (left and right figures, respectively).
Note: Red areas mark the highest signal strength, and blue areas indicate insufficient signal for communication.

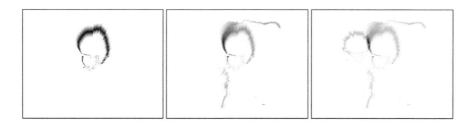

Fig. 6.8 Two-dimensional probability maps showing the most probable position of a tag.
Note: The left figure demonstrates the case in which signals from one reader is received, the middle refers to signals from two readers, and the right figure to signals from three readers. Darker shades mark areas with higher probability of finding the object.

as the ones shown in Figure 6.8. The far left image shows a probability map with one RFID tag in the cluster. The dark area marks the highest probability of finding the tagged object. The image in the middle shows the overlap of probability maps when two RFID readers are present and detect the tag's beacon signal. As is visible, there is a large reduction in the highest probability area in this case. In the far right image, the dark area is further narrowed due to the presence of a third RFID reader in the set.

A higher positioning accuracy can also be achieved through the time averaging of the signal. The image to the left in Figure 6.9 shows a probability map generated from data in a single simulation cycle. By overlapping data from consecutive cycles (as shown in the image to the right), the accuracy of the localization process is increased due to the reduced impact of fast fading. After ten or more cycles the fast fading has but a small impact on the measurements and as a result, positioning becomes very accurate.

The final probability map is then overlapped with the environment map: the highest probability area is specifically marked in red as the probable position. The administrator/user can change which object they wish to track by clicking on the relevant radio button in the panel marked "select tracked object" in the master board window.

The implementation of the positioning algorithm, as it would be displayed to users, is shown in Figure 6.10. The environment is shown on the image to the left. The middle image shows the probability map overlaid on the home environment. The yellow colors represent an area where the resident has a high probability of being located. Concentric red circles show the calculated

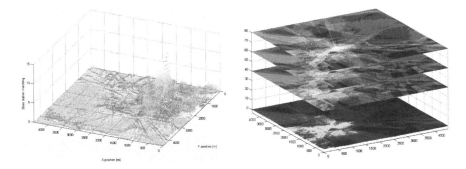

Fig. 6.9 Probability map representing a search for signals (left) and probability map comprising the time averaging of consecutive probability maps (right).

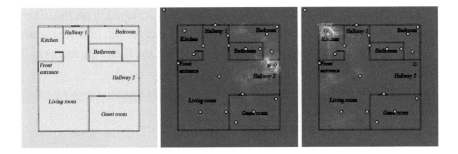

Fig. 6.10 Resident's home (left), probability maps and calculated positions for the resident (middle) and for tracked medication (right).
Note: The yellow colour represents a higher probability of finding the object. The calculated position is marked by red concentric circles. Green dots indicate resident movement hot-spots, and the red dot represents the exact position of the resident.

position of the resident and the red dot represents the actual position of the resident. Performing a similar query for the location of medication gives results shown in the image to the right. In this particular case, the medication is to be found in the kitchen area. The green dots on the map represent resident movement hotspots used for the simulation of the resident's movement throughout the house.

The master board (as shown in Figure 6.1) will typically be used and manipulated by a system administrator. A simple browser-based web version will be made available to the resident and the caregiver. This end-user interface would not provide control over the sensors and reader placement, but it would provide access to historical data, information on the current state of the timers and sensors. The resident would also have access to the localization routine for querying lost objects.

6.3.4   Simulation Results

As part of the research work prior to the publication of this book, a user-controlled simulation of the AGE@HOME software was run. The simulation in operation is displayed in Figure 6.11. This figure shows all active timers, the current state of all controls, as well as the relevant data in the history log. The current system time is shown in the system power panel on the upper left. Timers that are running appear in green in the appropriate boxes (e.g., the timers relating to no motion, the timer relating to the fridge, and the resident in timer). Their names and values are also displayed in the list of active timers. Those sensors that have been triggered, and are the basis for sending reports, are indicated by a yellow background (e.g., the RFID tag readers). The

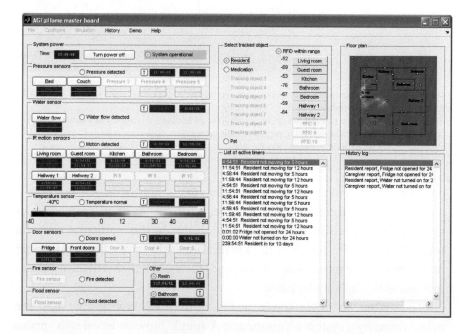

Fig. 6.11 AGE@HOME master board during random simulation.

Note: Currently active sensors are shown in yellow. Active timers have their respective displays counting down the remaining time. Active timer list updates every simulation tick. The history log contains all previously generated reports.

localization map at the top right of the interface shows the current actual (red dot) and calculated (concentric red circles) position of the resident. The areas indicated in yellow have a higher probability of finding the tracked object in question (in this case, the resident's RFID bracelet). The history log lists the reports that have been sent, the identity of the sensor that triggered them, and to whom they were sent (caregiver, resident or emergency services).

The AGE@HOME system is intended to provide history logs to end-users of observed data, through the use of the "History" menu. Recorded data for a bed sensor and associated timers during a 48-hour period is given in Figure 6.12. The blue line represents data from sensors. In this case, the bed sensor provided active feedback for a period of nearly 30 hours, indicating that the resident did not leave their bed during that time (this is indicated by the green line at the bottom of the graph).

The two timers associated with the bed sensor (i.e., a 12 hour timer 1, plotted in green, and a 24 hours timer 2), plotted in red have a time gap associated to them. This allows the timers to continue running for a short

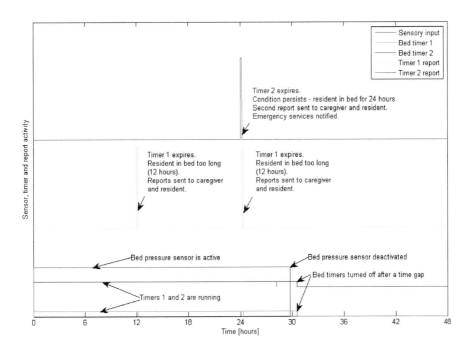

Fig. 6.12 Bed sensor, timer and report activity over a 48-hour period.

period of time, even though the sensor in question stopped sending any data. This time gap is used to avoid false reporting that might occur due to lost packets from packet collisions, other interference or fading. For the purposes of demonstration, the time gap in this example was increased from a default 1-minute gap to a 50-minute gap. The blue line in Figure 6.12 represents reporting activity for the first timer: reports sent to resident and caregiver after 12 hours and again after 24 hours). The magenta line represents reporting activity for the second associated timer: reports sent to resident, caregiver but also emergency services after the resident spends 24 hours in bed.

A history log of the timer and sensor activity for the whole system over a period of one hour is shown in Figure 6.13. This example illustrates a sample case, in which the resident spent too much time in bed. In this case, the second timer sent notification to both caregiver and emergency services. We can then assume that the resident was taken to hospital. In this scenario, the caregiver stayed behind to pick up a few personal items, and as a result, sensors show his/her movement and activity in the home, following the exit of the resident. The caregiver would have to disable any intruder alarms in such a case.

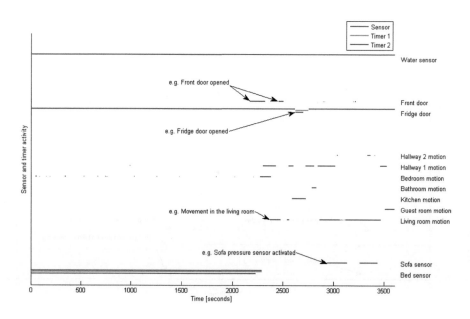

Fig. 6.13 Sensor and timer activity over a one-hour period.

The elapsed time in seconds is shown on the x axis, and the sensor activity is represented by blue spikes on the y axis. Each active sensor or timer is represented by a line and the name of the sensor is displayed on the right side of the graph. The red lines represent the running states of the first timer, and the green lines represent the running states of the second timers associated with each sensor. For the purposes of clarity, some events are further explained by text on the graph itself. For example, it can be seen how the opening of fridge doors (through data from the fridge door sensor) stops the second timer that is counting how long the fridge was not used (red line). It also starts the first timer which is responsible for monitoring whether the fridge door was left open for too long (green line). If no activity has been detected by the fridge door sensor for a longer time frame than the associated time gap duration, the first timer is stopped and reset, and the second timer begins to run.

The data as set out above was obtained using a time-accelerated simulation. An Intel Pentium M 1.86 GHz platform was able to perform the simulations three to four times faster than if they were done in real time: as a result, sensors and graphical interfaces were effectively updated many times per second. In a real-world scenario, most sensors and RFID tag readers operate with a two-second reporting interval. This signifies that the interface does not need to be updated more than once every two seconds. A processor running five times slower from the one used should be able to process all the data in a timely manner.

It is to be noted that the historical data as shown in Figures 6.12 and 6.13 can be made available to the end-users and health professionals in a different format, and covering longer periods of time. The implementation of behavior monitoring algorithms for specific purposes is made possible as a result. For instance, the number of hours the resident spends in bed each day can be monitored, and so, too, can the number of times a resident might have forgotten to turn off the water or close the fridge door. In this manner, caregivers or health professionals can keep a check on whether a resident is becoming more forgetful or following their daily routine as usual.

The simulation exercise successfully demonstrated that the decision logic for AGE@HOME, as first conceived in SDL, is applicable to a real-life scenario in the home. The random nature of the simulation took into account the possibility of false readings and lost packages. It was observed that the software performed the role for which it was intended, and matched the required parameters of the system as laid out in the initial concept, i.e., acting as a home behavioral and environmental monitoring system.

The simulation tests on the positioning accuracy with average environment noise, low number of rays in the ray-tracing algorithm, and no filtering, have shown a standard deviation error of positioning around 1.4 meters. With time averaging and basic filtering, this error was reduced to 0.25 meters for a 12×12 meters indoor environment. This is more than sufficient for indoor use. The request was processed within a tenth of a second on a P4 1.8 GHz laptop.

The AGE@Home software is therefore ready to be connected to a working laboratory for testing under real working conditions, with actual sensor input. The decision support system for AGE@HOME may be implemented on any suitable device in the home that supports the wireless interfaces to the RFID readers that relay the signals to the decision support system. This might be a home computer, a specialized controller or be integrated in a wireless router or home network gateway. Current alarm systems usually have a dedicated standalone controller that has an interface to wide area communication systems (mobile network, telephone network or Internet). Integrating the decision support system with a home "media center" would have the advantage of using a common user interface, such as a television (and remote control), as well as a large screen display.

6.4   On-demand Data and Historical Trends

An important benefit of the AGE@HOME system is that it can accumulate data collected from sensors and tags unobtrusively. This enables the provision of historical information on-demand to residents and/or their caregivers about significant behavioral trends, such as a decreasing number of hours spent sleeping and/or limited excursions outside the home. In this way, analysis of the overall medical condition of a resident is enhanced through an understanding of daily habits. Historical data can be limited in availability to the elderly resident if desired, but it may also prove useful for caregivers. Environmental trends such as water usage, temperature levels and so on can also be analyzed.

The following historical behavioral trends could be presented to the user on demand:

1. Time spent laying in bed per 24-hour period.
2. Time spent laying in bed between 11 pm and 10 am every day.

3. Time spent laying in bed between 12 pm and 8 pm every day.
4. Number of times resident leaves home per 24-hour period.
5. Number of times resident leaves home per 7-day period and 28-day period.
6. Number of times resident checks for lost items per 24-hour period.

Data may be obtained via a web-based application on a personal computer or television set. In the present case, a basic web application was developed using Adobe's Dreamweaver program. The data in the historical logs as well as the localization functions was directly obtained from the computer simulation in real-time. Figures 6.14 and 6.15 provide screenshots of the basic AGE@HOME web application. In Figure 6.14, an "I'm ok" button enables a resident to send SMS to advise caregivers that they are all right, after a report has been sent out. Figure 6.15 shows what might be displayed to a resident when he or she clicks on an item of importance to locate it, e.g., their own ID bracelet or a pet. In this interface, text can also be added to indicate the location

Fig. 6.14 Basic AGE@HOME application home page.

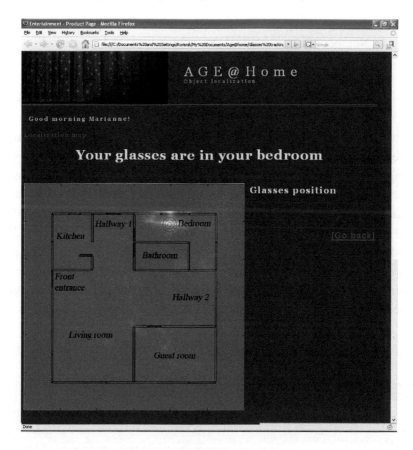

Fig. 6.15 AGE@HOME localization map.

of an item, in addition to providing a graphical image. At the implementation stage, a more sophisticated localization display mechanism will be required for elderly users.

6.5 Implications and Challenges for Developers, Policy-makers, and Users

This section describes some of the important implications and challenges of implementing the AGE@HOME platform, not only for technology developers, but also for policy-makers and users.

6.5.1 Design Challenges

The main design challenges for the AGE@HOME platform are related to system integration, choice of wireless networking technology, and data security. Currently, there are only very few sensors available on the market with integrated tag. The use of RFID to transmit sensor data is not common, and is certainly rare in the home — it is mainly in use today in industrial contexts. The combination is an interesting one from a technical standpoint, and further research should be conducted on the possibility of making the tag and sensor intelligent, to permit adjustable parameters within the sensor (e.g., this would be particularly useful for temperature sensors and biomedical sensors).

Currently, wireless LANs (e.g., Wi-Fi or IEEE 802.11b) are widely used in the home for communication between computers and home network gateway routers and Bluetooth is often used for shorter range communications between such devices as a mobile phone and a computer and for remote control applications. There are many other technologies available, most of which are proprietary with some form of industrial alliance organization to promote their adoption. The use of ZigBee is an interesting alternative: based on the IEEE Wireless Personal Area Network (WPAN) specification 802.15.4, it is a promising standards-based approach that is expected to be particularly beneficial for health care devices. As previously mentioned, it is proposed that RFID be used to transmit data from sensors that are equipped with RFID tags, and that WLAN be used for data transmission between RFID readers and the central controller. It is also proposed that a wireless technology such as Bluetooth be employed for the user control interface. In a laboratory environment, it may be interesting to experiment with different networking technologies for data transmission between readers.

It is important that widely adopted interface specifications are used to ensure interoperability of the various devices within the home. The various wireless technologies must work, of course, so as not to interfere with each other or with wider area network communications. This will be an important integration challenge when transferring the theoretical concept presented here into an experimental context.

Currently, many entertainment functions are being integrated into home media centers through the use of home servers. The integration of various functions on a home server will avoid the presence of a large number of independent systems with their own user interfaces and remote controls. A standard user interface and large screen display will become more easily available. However, this may be difficult to achieve as equipment vendors typically compete on the basis of novel user interfaces and applications. Conflicts between proprietary systems and open systems, competition and collaboration persist. However, it is vital at this time for industry to step up integration and interoperability efforts in order to achieve the basic capabilities of the digital home, and perhaps more strategically, a wider user base.

With respect to data security, user access controls to the system should be put into place, and data collected by the decision support system should be adequately protected. An authentication system should be used in addition to any encryption for data transmitted over the internet to a web-based application.

The provision of the service is yet another challenge. Independent living applications may be provided by a network service provider or health service authority. Alternatively, these services may also be implemented as private standalone systems. Managed systems will naturally require higher levels of security and data protection. But as RFID tags hit the mass market, their use will have to be more controlled, and mechanisms to temporarily or permanently disable tag data transfer will have to be put into place.

Finally, the user and management interfaces must be straightforward and intuitively easy to grasp. Much work is needed in order to design systems that are easy to use and that target populations such as the elderly, who are often challenged in terms of sensory perception (vision, hearing etc.). The overall system should also be made affordable enough to stimulate mass adoption. This is why the use of traditional technology is suggested here, such as motion sensors in use with alarm systems, and wireless LAN, which is already widely used in the home.

6.5.2   User Choice, User Control

It is vital for any system that is based upon the monitoring of individual behavior patterns, that it be accessible and controllable by the users concerned. If a user wishes to opt-out of a particular service or sub-service, a simple procedure should be in place. In the case of AGE@HOME, although the caregiver is given access to resident data, the ultimate control of that data rests in the hands of the resident. It is assumed that residents using the system are able and fit to live on their own. However, this raises an important dilemma: should the information generated by the system be used to base decisions about the need for nursing home care on a case-by-case basis? This can have positive and negative implications: it can protect individuals in need of care, but can also be subject to abuse, depending on the intentions of those wishing to place residents in nursing home care. It must be recalled, that the overall objective of the system is to prolong as much as possible the period for which the elderly can live independently. Thus, it is imperative that elderly people be given control of this data and its distribution. It is intended that AGE@HOME data not be made available outside the closed user group of resident and caregiver, and only in exceptional circumstances (with the permission of the resident) shared with health care professionals. The key advantage of AGE@HOME is that it obviates the use of surveillance cameras and may therefore be more easily acceptable to users.

Users must also be given the appropriate knowledge and know-how to use the system. It goes without saying that a certain minimum level of education and training will be required. Many argue that this is necessary not only for emerging technologies, but also for simple daily internet browsing and chatting.

Potential barriers to adoption must be taken into account. The possible reluctance of elderly, handicapped or technology-averse individuals is a significant factor. Pro-active measures must be taken when introducing these types of technologies in the home, and various "soft" factors affecting adoption should be carefully considered.

6.5.3 Policy-making in the Public Interest

It is important for the public sector to recognize the central role played by policy-making in innovation for emerging technologies. In the case of RFID, it is incumbent upon governments and regulators to create mechanisms raising awareness about the technology, its benefits and pitfalls. They also have a responsibility to ensure that users are aware when and in what context RFID is being used. There may be a need to reconsider data protection and privacy legislation in light of current technological developments.

Governments also play an important role in providing necessary incentives and funding for research and innovation in the field of technology for the elderly, data security, and health protection and prevention. They are also instrumental in promoting standardization efforts and extending public access to networks, information, and new technologies.

Considerable thought must be given to the governance of digital resources. A domain name system may soon manage not only the web sites of today but also the web sites of tomorrow, which may refer to the life of individual items or geographical spaces. As things are increasingly equipped with individual identifiers, these identifiers will become strategic marketing tools, even separate commodities. In this context, the mechanisms by which this global database of identifiers will be managed have yet to be determined. Who should govern tomorrow's internet: governments, industry, individuals, or international organizations? The outcome is uncertain, but one thing is clear — our approach should be democratic, transparent, multilateral and humanistic, particularly as the commercial importance of the internet continues to grow rapidly.

It is vital for governments to continuously re-examine their own objectives and measure the efficiency of their policies. In a global era, one in which business interests are often paramount, it is increasingly important to foster public–private partnerships and ensure that industry collaborates with government in its policy-making process.

Finally, policies need to reflect the power imbalance between policy-makers and policy subjects. As a group, the elderly are not in a position to put forward their interests as effectively as other lobbying groups, and certainly not as well as industry. They are less organized, have fewer resources, and are typically considered a minority voice in society. Ageing is often viewed as a long-term concern, and can therefore be marginalized vis-à-vis shorter economic objectives. Governments should be proactive in gaining a better understanding of the factors at play (e.g., through solicitation of views, market analysis, demographic mapping, and so on). They need to devote adequate attention to the development and institution of policies that will eventually benefit a growing sector of the population.

7

Living Digital

We don't stop playing because we grow old, We grow old because we stop playing.

— George Bernard Shaw

This chapter considers the growing scope of networks and the evolution of life in the digital age, particularly for the elderly. It begins by outlining the case of Japan, one of the leading countries in ICT innovation, and coincidentally one which faces a severe ageing challenge of its own. It concludes with an examination of some of the socio-ethical perspectives needed for gentler and more effective technology design.

7.1 Case Study: Ubiquitously Digital Japan

The position of Japan as a world leader in the field of information and communication technology (ICT) is widely acknowledged. The perception that the Japanese are a highly technophile people is confirmed time and time again as they are seen sporting the latest gadgets. They are at the cutting edge in the field of robotics, home electronics, product packaging, and device miniaturization [1]. Japanese public policy is squarely focused on the notion that ICT is instrumental in stimulating social and economic structures. Addressing the ageing phenomenon is no exception. For this reason, Japan is an interesting example to consider in the context of the present work, particularly for its vision of an all-inclusive ubiquitous network society and lifestyles.

7.1.1 Ageing in Japan

In 2006, Japan had a total population of 127.8 million, making it the tenth most populated country in the world [2]. The pace at which its population

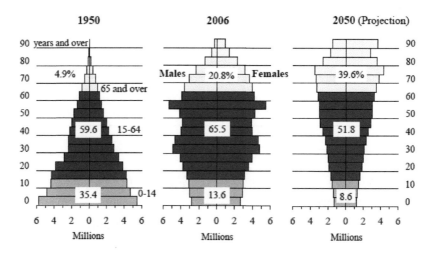

Fig. 7.1 Japan's population pyramid (1950, 2006, and 2050).
Source: Statistics Bureau, MIC, Ministry of Health, Labour and Welfare [2].

is of ageing is one of the fastest in the world, second only to Italy (see Figure 2.2). In 2006, the number of Japanese over the age of 65 reached an all-time high, at 26.6 million, or 20.8 percent of the total population (Figure 7.1). And this is set to increase yet. In 2007, there were 27.5 million people aged over 65, of which 12.7 million were over the age of 74. This is expected to rise to 25 percent in 2013, almost 34 percent in 2035 [3], and all but 40 percent in 2050 (Figure 7.1). The Japanese are enjoying longer lives, due to, *inter alia*, healthy diets and advances in medical care. But longevity implies increased risk of illness or injury. In 2005, the number of the elderly in need of nursing-care stood at 4.17 million, up from 2.8 million in 2000, corresponding to 16.6 percent of the total elderly population. This number is expected to rise to 5.2 million by 2025 [4].

In response to these trends, the Japanese government has identified the consequences of ageing as one of its top priorities, introducing various policy packages and legislative reforms. More specifically, these policies place a strong focus on the opportunities that emerging technologies provide to improve the quality of life of elderly persons and those with disabilities.

7.1.2   Policies for ICT in Daily Life: From e-Japan to u-Japan

The government considers the role of ICTs as vital to social and economic development. This has been confirmed by statistics — by government estimates, ICTs have contributed as much as 40 percent to national GDP (Gross Domestic Product) in 2006 [5]. Since 2001, Japan has been establishing a series of national policies with the aim of making Japan the most advanced nation in the world in respect of ICT. In 2001, the e-Japan strategy was launched, with a policy priority program that included five policy areas, namely infrastructure, human resources, e-commerce, e-government, and network security [6]. The e-Japan strategy II was adopted in 2003, with clear targets by 2006 [7]. This phase focused not only on infrastructure but also on usage and user adoption. Seven key areas were designated for the promotion of the use of ICTs, viz., medical care, food, living, small business financing, intellectuality, employment/work, and government services.

After having addressed infrastructure and usage in its *e*-Japan strategy, the government devised a new *u*-Japan concept in 2004, and launched it in 2005 [8]. The evolution from an e-Japan to a u-Japan strategy is set out in Figure 7.2. Based on a vision of networks available anytime, anywhere, by anyone and anything, the u-Japan strategy contains four different policy packages. Each of these have a number of focus areas, including the networking of real objects, through technologies like RFID, and the promotion of universal design for easier accessibility for the elderly and handicapped (Table 7.1).

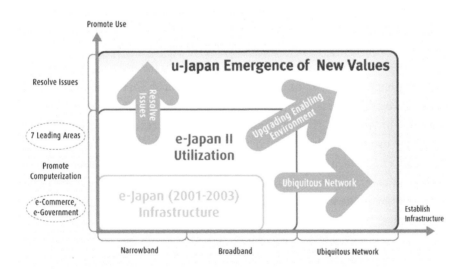

Fig. 7.2 From e-Japan to u-Japan.
Source: Ministry of Internal Affairs and Communications, Japan [9].

Table 7.1 U-Japan policy packages.

Policy packages	Areas
1. Development of ubiquitous networks	• developing seamless access environment for both fixed and wireless networks • nationwide establishment of broadband infrastructure • networking real objects • developing infrastructure for network collaboration
2. Advanced usage of ICT	• sophisticated social system reform by ICT • promoting creation, trading, and use of content • promoting universal design • ICT human resource development
3. Upgrading enabling environment	• promoting 21 strategies for ICT safety and security • establishing the Ubiquitous Network Society Charter
4. International and technology strategies	• promotion of policies not only for domestic society but also for international markets and networks (international) • strategic promotion of R&D and standardization in priority areas, and to strengthen international competitiveness through innovation (technology strategy)

Source: Adapted from Ministry of Internal Affairs and Communications, Japan [8].

One of the aims of the u-Japan strategy is that by 2010, 80 percent of the Japanese population should believe that ICTs are useful in solving problems and issues, for instance health and safety. Another aim is for 80 percent of Japanese people to feel comfortable with ICT. The policy does state that these objectives may require a certain modification in "systems, customs, and habits that may not readily accommodate ICT" [10].

In this context, the u-Japan strategy is squarely based on user-centric innovation policies. The "u" of u-Japan refers to four different elements: ubiquitous, universal and user-friendly, user-oriented, and unique, informally known as "4U" or "for you" (Figure 7.3). Under the universal and user-friendly concept, the use of ICTs by the aged and disabled is emphasized.

The vision of a ubiquitous network is one that enables users to safely, and with ease, use network terminals and devices, and access digital content without having to actively think about using them. The communication must be seamless between people and devices, between devices and things, between people and things, and between things and other things, breaking down all preexisting physical and virtual boundaries [12]. The number of application areas for the ubiquitous network society is limited only by human imagination, and can range from entertainment and disaster relief to health care and shopping.

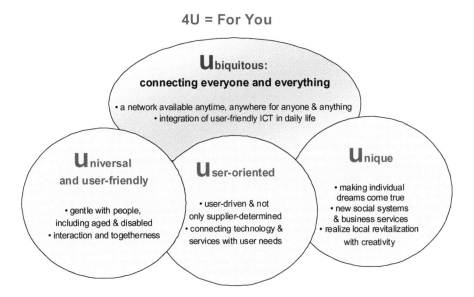

Fig. 7.3 The user-centred principles of u-Japan.
Source: Adapted from Ministry of Internal Affairs and Communications, Japan [11].

7.1.3   RFID in the Network

Japan has been a leader in the adoption and deployment of RFID systems. It was one of the first countries to introduce RFID retail shopping, known as the R-Click service, in its Roppongi Hills area [1]. Microchip networking technologies are an important part of the government's research and development agenda. The u-Japan policy underscores the importance of developing RFID tags to be combined with sensor networks and network robots (Figure 7.4). There is also a public policy focus on standardization and networking of information appliances through the use of an expanded IPv6 addressing system. One of the greatest challenges is to ensure that communications are not only made possible between heterogeneous devices but also that collaborative data is collected and distributed when needed, to the right destination, in the right context, and in a secure and rapid manner.

   A good example of an early safety application for RFID in use in Japan is the monitoring of children going to and from school. In some schools, pupils now wear active RFID tags in their backpacks, so that their entry into school property can be detected through readers at the gate. This information

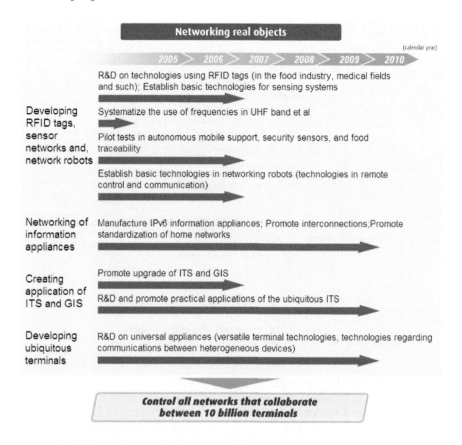

Fig. 7.4 Japanese policy timeline for networking objects under the u-Japan strategy.
Source: Ministry of Internal Affairs and Communications, Japan [9].

is stored in the school's attendance database. A text message can also be sent to concerned parents as to the time of arrival and departure from school [13]. Similar uses for RFID are being developed to track entrances and exits in the home (see Section 7.1.4).

RFID can also help the blind and elderly navigate. Projects in Japan are exploring blind navigation systems using RFID readers in walking sticks and tags in public spaces that alert of obstacles being approached. Traffic lights can also be equipped with readers that detect the elderly and handicapped, tagged with RFID, and ensure that the traffic lights do not change until they have finished crossing the street [14].

The use of RFID is viewed in Japan as integral to public safety and convenience. In December 2006, a trial project called the Tokyo Ubiquitous Network Project was launched. The famous Ginza neighborhood in Tokyo was equipped with a wide-area RFID system of readers and 10,000 RFID tags. The readers and tags provide location-based information to visitors equipped with specialized reader devices. The code sent by each tag is stored on the Internet, and information is communicated via a wireless LAN connection [15]. Plans to expand the smart RFID system to other parts of Tokyo are underway.

With the spread of RFID applications in the country, Japan's government has a number of initiatives in place to deal with privacy and data protection [16]. It aims to ensure that users have full control on the information related to their location, purchasing habits, and so on. As early as 2004, the government issued specific "Guidelines for Privacy Protection with Regard to RFID Tags" [17].

7.1.4 Building the Digital Home

Japan's u-Japan policy is emphasizing the networking of public spaces as well as private spaces. Research and development on the "ubiquitous home" is ongoing and, like many other countries, Japan is facing similar challenges with the incremental and fragmented nature of digital home adoption. A number of specific home security services, content services and such are available on the market but each is sold by different service providers. As a result, users face confusion and high costs when trying to avail themselves of a networked digital environment at home.

For instance, there are commercial home security systems using RFID readers in mobile phones. A new housing development northeast of Tokyo has included Japanese operator NTT DoCoMo's wallet phone to ensure that access to the residence is only granted to someone carrying a phone with a pre-registered RFID tag. The mobile phones can also connect to a home server allowing residents to switch on lights, heaters or take photographs of unexpected callers. Email alerts can be sent to residents of anyone entering or leaving the house using a phone key. Although the system is costly, the set-up costs are borne by the developer, with residents paying a monthly service fee [18]. Many argue, and rightly so, that such services, together with those that track children and shoppers, must be subject to stringent privacy protection mechanisms.

Other digital home systems focus on entertainment and content and still others, using biosensors, track the health and safety of residents. Domestic robots are on the rise, with Japan home to almost half of the world's industrial robots [19]. There is a strong commitment in both public policy and R&D circles in Japan to finally integrate the potential of emerging technologies, including RFID, robotics and sensors, to create the truly integrated "ubiquitous home" for the convenience and safety of young, elderly or disabled citizens.

7.1.5 Changing Social Lifestyles

Technology has considerably changed social lifestyles and Japan is no exception. More than 30 percent of Japanese believe that their social life has changed over the last two years, and more than half of them attribute changes in shopping and entertainment habits to the advent of the internet (Figure 7.5). Patterns of network usage are clearly affecting lifestyles in Japan. With regard to the elderly, those over the age of 60 increased their use of the internet significantly between year-end 2003 and 2006 [20]. Although they have not reached the level of usage of their younger counterparts, it is certainly important to note that as time goes on, the elderly are likely to become even more important users of the internet. As a result, products and services should be increasingly geared toward their needs and requirements.

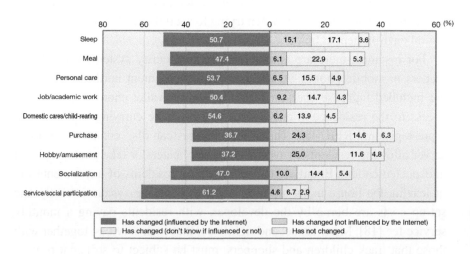

Fig. 7.5 Changes in daily activities over the past 1–2 years and the impact of the internet.
Source: Ministry of Internal Affairs and Communications, Japan [20].

Japan provides a good example of the potential of ICTs to enhance daily life in a timely and comprehensive manner. Policies and strategies on emerging technologies are being actively combined with social policy on public safety and welfare, in particular understanding and addressing the needs of the elderly and the handicapped.

7.2 Networking Things

Today's innovations in automatic data capture and identification will fuel the coming of tomorrow's so-called "internet of things" [21], in which everyday objects dynamically interact with a global network of networks, transmitting information and knowledge seamlessly and securely where, when, and how it is needed. In this context, RFID and related technologies are only the beginning of a paradigm shift in network communications and computing.

The vision of the "internet of things" will require a new approach to network architecture, which must not only be robust, but infinitely scalable. A balance needs to be struck between distributed and centralized networks, between placing intelligence in the network and at its edges. Global standards enabling interconnection and interoperability are of course vital. As the network grows exponentially, critical resources such as the radio spectrum will need new management techniques (both technical and regulatory), enabling dynamic access and greater possibilities for sharing and re-use. Power management issues will also be crucial, as current limitations on battery life cannot keep up with demand for innovation in wireless services and applications. A better understanding of social trends, both global and local, will ensure that both innovators and service providers meet user requirements for simplicity, affordability, and "sociability." There is a need to develop new interfaces and protocols for machine-to-machine communications and human-to-machine communications, and to convert complex systems into simple plug and play solutions. And as systems become increasingly complex and expansive, the need for network integrity and security will become even more acute.

These issues need to be resolved for the new network to be truly successful. The ideal network is one that is able to learn, adapt, and inform. We might imagine a world in which environments are context-aware and cognitive, continuously able to adapt to changing circumstances and to short or long-term variations in user preferences or requirements. At the heart of this vision lie two important developments: today's web 2.0 and tomorrow's internet of things.

7.2.1   All About Identity

Today's internet is made up of a series of interconnected computer networks that communicate with the help of technical protocols governing the exchange of data. Aptly named a "network of networks," it functions on the basis of an addressing system known as the domain name system (DNS). The DNS is hidden to users, but helps them navigate the Internet. Every computer on the Internet has a unique address (an Internet Protocol address or IP address) made up of a string of numbers, e.g., 138.23.65.127. Naturally, computers communicate these numbers in binary form. The conversion of these numbers to a string of letters (e.g. www.cnn.com), known as resolvability, makes it easier for users to find their way around the internet (Figure 7.6). Today's IPv4 defines a 32-bit address, and as a result, only 232 (4,294,967,296) IPv4 addresses are

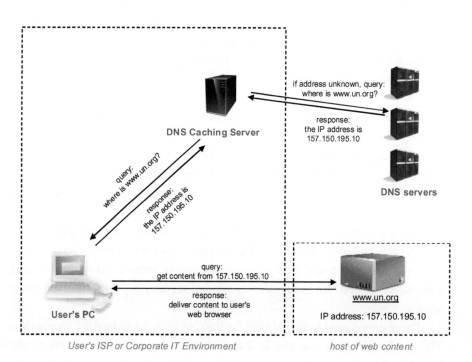

Fig. 7.6 Identifying domain names.

available, corresponding to 4.29 billion possible combinations. This is less than the world's present population. Ongoing work IPv6 hopes to rectify this scarcity, by defining a 128-bit address space. An IP address, however, can only go so far. It acts more like a locator than it does as an identifier of a particular device. However, to a certain extent, today's internet is the beginning of a new approach to identification and identity. In the new always-on digital lifestyle, the way identity is perceived and communicated is changing. People are now contactable via a nongeographic email address, and use this to communicate, transact, and identify themselves. They also use personal web sites, pages on social networking sites, and gaming identities. In the past, people were primarily identified by the name of their village, combined with the name of their family. As a result, it was fairly difficult to change identities or socio-economic status. Where, and to whom, an individual was born was largely determinative of their identity, where they called home, and their source of livelihood. With the modern era came the possibility of participating in different social circles and the possibility of improving socio-economic conditions. In today's world, the individual has much more choice and is able to participate in a greater number of social networks, sometimes disparate ones. In the digital space, individuals may identify themselves differently, depending on context or choice. The creation of multiple selves can of course enrich and empower, but it can also dilute identity and self-governance, and carries with it the potential for misrepresentation, fraud, and criminal activity.

The future network — a network of things as described above — will attribute an address or identifier to everyday items so that smaller and smaller objects become networked (i.e., enabled by further evolutions of IPv6, electronic product codes, *u*codes, etc.). As a consequence, the multiple nature of identities would not only apply to humans, but also to machines, devices, networks, objects, and resources [22]. Ongoing and concurrent developments in chemistry, genetics, biometrics, robotics, and nanotechnology will further expand the scope and impact of identification processes and of identity as a whole. Before that happens, however, we will have to examine on a much higher level the evolution of human identity and its changing patterns. In order to do so, we will have to evaluate how today's technologies and applications are changing the way we communicate, learn, socialize, play, create, govern, and transact. In particular, the mechanisms by which we represent ourselves to others (people, devices, institutions etc...) online will have to be revisited.

At the moment, no global identity management scheme exists, yet identity is a core facet of using online services. The fragmentation of online identities is a growing concern, with most internet users holding a variety of users names and passwords, depending on context and preferences. This is not only becoming unmanageable for the users themselves, but it also results in the lack of accountability. At the same time, the notion of anonymity must be preserved for situations in which full disclosure of identity is not required, and may even prevent users from making full use of services, and full use of their rights under the law. One of the biggest advantages of the digital age is that it is has given a chance to those who have not had a voice in the past. At the same time, one of its greatest pitfalls is that it can make users more vulnerable to fraud and other forms of cybercrime stemming from misrepresentation. The solution lies in a system which is context-based, and favors the collection of a bare minimum of information for each situation. Such a system might be based on groups of "partial identities," depending on context, so that information on health is not shared with employers, or information about banking is not shared with hospitals, and so on. Each individual may have a number of data points that defines them as unique, but not all are needed to transact in any particular context, and allowances must be made for user choice (Figure 7.7). At the moment, there are increasing calls at the international level to create a global system for identity management [23], but much work remains to be done in this area.

7.2.2   Web 2.0 and the Internet of Things

In exploring how technologies have impacted the way we learn, create and interact, Web 2.0 [24] stands out as a striking example. It constitutes a new approach to using the internet, and in particular the World Wide Web. It has emerged over the last few years and is much more participatory and collaborative in nature (Figure 7.8). It has been driven by innovation in applications and software that encourage creativity, information sharing, trust, and user collaboration. This appellation has now gained wide acceptance to indicate a departure from the previous incarnation of the web. In the early days, the web was principally a worldwide repository of information for passive consultation by users. Now it is more of an open and dynamic platform for user-generated content, socialization, entertainment, knowledge-sharing, and

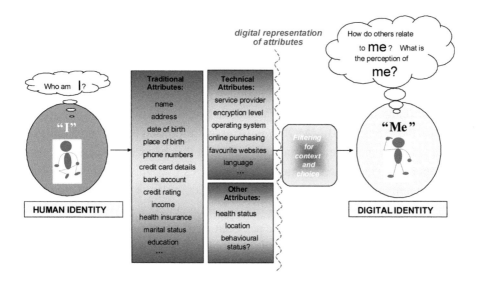

Fig. 7.7 Digital identity.
Source: Adapted from, ITU Internet Reports 2006: *digital.life* (ed. L. Srivastava) [22].

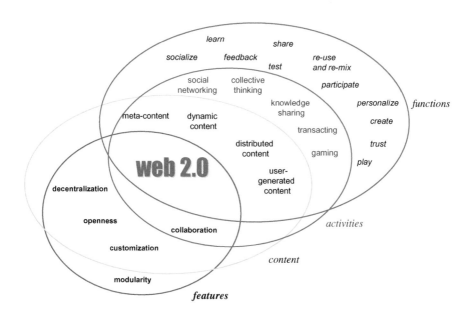

Fig. 7.8 The workings of Web 2.0.

collective thinking. A good example of how usage of the internet has changed with web 2.0 is the rise of applications such as Wikipedia [25] (and subsequently other "wikis"), which users can update and edit themselves, as opposed to the use of services such as Britannica online, which is limited to consultation. But the so-called Web 2.0 includes other developments, such as social networking sites (e.g., Bebo, Facebook, MySpace), media sharing applications like Flickr, and MMORPGs (Massive multiplayer online role-playing games). Another phenomenal appearance is that of virtual worlds, in which users can create representations of themselves ("avatars") and even become entrepreneurs, adding their own value to the complexity and sophistication of their online world (e.g., Second Life). With web 2.0, everyone is an internet celebrity. Instead of largely static personal web sites, users can turn to dynamic blogging to maintain and update their online presence. Individual blogs are now appearing in a majority of search engine results on topics ranging from movie/book reviews to disciplines such as philosophy or economics. The web feed format RSS (Really Simple Syndication) has been instrumental in giving users the possibility to publish frequently updated information and to keep up with web sites in an automated manner. Most of the larger web sites now leave room for comments by users. They also enable "social bookmarking" by users for storing, tagging, and sharing links: some of the more popular services include del.icio.us (http://del.icio.us/), reddit (http://reddit.com/), and digg (http://digg.com/).

Web 2.0 is at the heart of the new digital "meta-content" universe that is rapidly searchable and is perpetually updated. On Google Earth, for instance, users can upload pictures of specific locations and sights, and the program cross-links to relevant YouTube videos. This distributed dynamic content gives users a sense of community and ownership, and facilitates knowledge-sharing and collective thinking on an unprecedented scale.

The future network is envisioned as one in which each of the world's individual objects would, in a certain sense, have its own web page, providing information on its origin, date of manufacture, ownership, date of expiry, location, components, and so on. As web 2.0 meets this new internet of things, the provision of meta-links to content and photos of real physical locations, would be combined with processes enabling physical locations to provide meta-links to virtual content. In a museum, tags might link users through universally available devices, like mobile phones, to a wealth of online information, user comments, recommendations, and virtual tours. They may help the blind navigate through (and update) their urban environment by sending messages to their walking aid. Users could add information to tagged objects throughout their environment, for real-time uploading to the virtual web. This would create a veritable two-way nexus between the real and virtual worlds. Context-awareness would be further enabled by true service portability across networks, applications, and devices. The real world would be mapped in the virtual environment. Similarly, the virtual would be mapped in the real world, through a combination of the dynamic user-generated content of web 2.0 and the ubiquitous computing and environmentally aware capabilities of the internet of things. As a result, day-to-day objects and locations would not only become digital artefacts, but social artefacts, each capable of telling its own unique story.

Although the open, user-driven nature of web 2.0 is a considerable advance, yet its interpretation of the user's expectation and its understanding of the context are quite limited. The choice available to the user remains less than satisfactory. Nevertheless, it is a promising beginning, and one which, in combination with other developments such as the internet of things, may give us a glimpse of its future form.

7.2.3 User-Centric Virtual Mapping

The linking and mapping of the real and virtual worlds is a continuous and important long-term objective. It will enable us to fully leverage online content in the real world, and will also serve to augment virtual learning, interaction, and transaction. Such "linkability," however, should not lead to the "outing" of all anonymity online. The real and virtual worlds should be brought closer together not via an "invasion" of one by the other, but rather through exploiting the potential for enhancing day-to-day experiences online and offline. Avid users of the "Second Life" virtual world, for instance, may wish to keep their "second life" independent of their real life, or what has become known in such circles as their "first life." This should continue to be possible — after all, it is the internet's potential for anonymity that has given a platform to so many minority voices afraid of speaking out publicly, and also fuelled interactions between people who may have otherwise been excluded from usual social interaction. Nonetheless, mechanisms to partially identify individuals, if and when context dictates, should be put in place: full anonymity may be desirable in the case of surfing the contents of a bookstore, whereas greater "identifiabil-ity" may be required when purchasing property. In some cases, users may wish to enhance their real world experience at home, by using avatars to interact with friends far away, or by identifying themselves (and the knowledge they possess) to objects and devices in their environment. The real and the virtual may be linked but not inextricably merged. This will lead to utter confusion and the benefits to be realized through such an arrangement will be totally drowned in the resultant chaos. Nevertheless, anonymity and its associated benefits should maintain subject to justifiable controls.

Virtual mapping in a dynamic real-time manner will be the next step in the development of a context-aware future network. Just as the domain name sys-tem now "resolves" a string of numbers in order to locate content on the web, so too will an internet of "identities" resolve identities to navigate a universe of dynamic and perpetually updated meta-content. There are key challenges to overcome, however, and these must continue to be subject to periodical reconsideration as both technology and society evolve: from the user's per-spective, data privacy and information accuracy are key concerns, and from the information systems perspective, network/data security and reliability are vital. Common to both perspectives is the requirement for secure identity man-agement mechanisms for the various resources, things and people interacting in the digital space. Resolving identity will enable us to fully exploit the "i" in internet and the "i" in individual, to create a truly user-centric network — in other words, to bring about the necessary but radical transformation from a web 2.0 to a web i.0 (Figure 7.9).

A Web i.0 is centered on the user and their daily activities. These activities occur simultaneously and at different times. Either way there is real benefit in making them complementary. For instance, doing chores can be combined with gaming, shopping with learning and social networking. Taking medica-tion can be combined with providing user feedback. Caring for the elderly can be done while travelling or at the office. A web i.0 would retain not only the participatory and open aspects of web 2.0, but would also be greatly enhanced by a context-aware mapping of the user's environment and by the portable nature of that environment. It is the user and their identity that drives web i.0. This identity is based on conventional data points relating to static informa-tion (e.g., date of birth) as well as dynamic information (e.g., preferences,

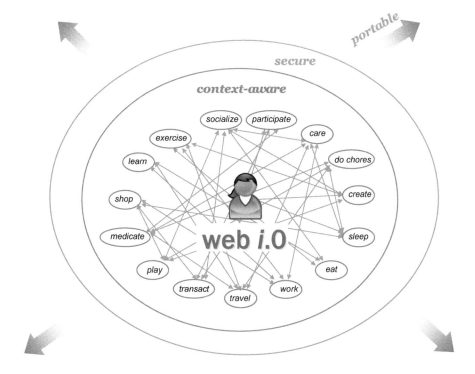

Fig. 7.9 The vision of web i.0.

location). These data points should be made secure and portable across inter-faces, devices and contexts, so that users benefit from the same services wher-ever they find themselves and this regardless of the platform they use. But the data transmitted in each case should be the minimum dictated by the govern-ing context. This should serve to protect the user and the data associated with them, keeping the intrinsic "i" of the user central and secure.

7.3 The Digital Aged

In light of the above, and given that today's young income-earners will reach the age of dependency at a time when the beginnings of web 2.0 will be but a distant memory, it is important to leverage and channel today's innovations to benefit the elderly in innovative and creative ways. Though elderly people have traditionally been laggards in technology adoption, this is not adequate rationale for limiting innovation development in this area, or limiting marketing efforts to younger age groups.

The elderly in a number of countries now make up for a large proportion of total internet use. The United Kingdom communications regulator released a report in 2007, which shows that web users over the age of 64 spend more time online than any other age group, averaging over 40 hours per week, though these users are predominantly male. Furthermore, with increased longevity, the cultural attitude to ageing is changing: the sixties are now the new fifties and retirement is a new beginning rather than an end to life. People have specific needs as they grow older, and these needs can represent both a business opportunity and a social opportunity. Home robots, for instance, as companions or caregivers, may be particularly well suited to the aged population, and as such, they may represent lead users of the technology.

As young users of today's technology get older, the question is how their attitudes to these technologies will change and how they will stay the same. The elderly today do not have memories of using the internet or the mobile phone when they were young, but what if they had never known a world without anywhere, anytime communications? How could those habits be channeled to assist them as they grow old in a society that is increasingly short of resources, from energy to health care? At the same time, although network interaction between everyday objects in our environment may seem like borderline science fiction to us, it may soon be a commonplace for our children. How will their habits follow them into adulthood and beyond? Will they continue to use Second Life or social networking sites, and if so, how will their attitudes change?

7.3.1 Community Building, Digital Sociability, and Accessibility

Although sites like MySpace and Facebook are the most well-known, there are a growing number of emerging social networking sites that focus on hobbies or specific interests, and that are targeting certain ethnicities, age groups, health problems, and so on. Some users of the mainstream networking sites are finding the communities too large to handle, and are choosing to target like-minded people through specialized sites.

One of the greatest benefits that new web 2.0 applications can offer is the sense of community and ownership. This becomes increasingly important with age. And as the number of elderly people rises, there is a sense of greater urgency to stimulate fulfilling and quality day-to-day living. Loneliness and social exclusion will be significant hurdles to overcome as single-person households become more commonplace. Finding other like-minded individuals with whom to socialize is a challenge for most, but more so for the aged. Online community building can be an important tool for the elderly to build and maintain relationships, to find support groups, entertain themselves, and continue lifelong learning.

Social networking

A spoof image of how young Facebook users may use the application when they get older has recently appeared on the web (Figure 7.10). Though not a real application, "pensionbook" as it is known does point to the growing need to target online social networking applications to an ageing population. Social networking has a number of inherent benefits for the elderly: it could enable them to more easily share tips, information and comments about healthcare products, keep in touch with loved ones or locate lost friends. Sharing common interests, photos and files would be facilitated, and the elderly may enjoy the possibility for real-time chatting and asynchronous messaging, beyond

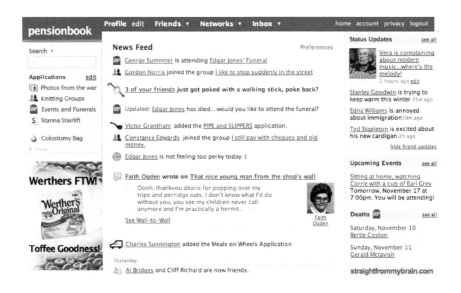

Fig. 7.10 Facebook for tomorrow's elderly?

Source: S. Wildish [30].

conventional email applications. From a healthcare perspective, perhaps one of the most important functions of social networking might be the creation of support groups about diseases and conditions such as Alzheimer's, diabetes, visual impairment, and osteoporosis. At the moment, the majority of these sites are geared toward younger users, although the aged, too, represent a viable market to aim at. For naturally, they have a good deal more time to spend on-line than their younger counterparts. Additionally, the benefits of always-on network and information availability can be particularly beneficial for isolated and vulnerable groups.

Knowledge sharing

The combination of social networking with information portals could offer real-time health information, not only for health care professionals and care-givers, but also for patients and the elderly. A good example of a healthcare web site that has taken on board the web 2.0 approach is ENURGI [26]. The aim of ENURGI is to connect elderly and physically challenged individuals with health professionals and caregivers. Users and suppliers of services can set up their own profile easily and conduct needs-based searches. Sites such as "TauMed" [27], "Revolution health" [28], and "iMedix" [29], provide vir-tual health communities for people to share information and opinions about consumer health and consult databases of health-related information.

The website iGuard [31] enables users to search for information on med-ications (prescriptions, over-the-counter medication, and nutritional supple-ments) and to share feedback and opinions. The system routinely checks the medication that users list and alerts users of any contra-indications, safety updates, or recalls. Its general safety checks and alerts use a system of risk rating, as set out in Table 7.2. Doctors, too, can be alerted if important safety information about medications is made available.

Gaming

The online gaming world has customarily been dominated by young male users. Recently, the gaming industry has realized an untapped opportunity: the female gamer market. It has not, however, taken sufficient note of elderly users as an important market segment, particularly, as mentioned above, given the greater amount of spare time the elderly have available for entertainment.

Gaming for the elderly could prove beneficial by keeping both mind and body fit. Table 7.3 sets out some of the potential benefits of gaming. On

Table 7.2 iGuard Safety check risk ratings for users.

Low Risk
Suitable for Widespread Use
These drugs have been widely used for a long period of time — which means that their safety profile is well known — and they are associated with a low risk of serious side effects.

General Risk
Use under Normal Care of a Doctor
These drugs have been widely used for a long period of time — which means that their safety profile is well known — and are associated with a low risk of serious side effects when used according to their directions under the ongoing supervision of a doctor.

BLUE

Guarded
Be on the Lookout for Safety Events
Most new drugs fall into this category. They have been shown to have a low risk of serious side effects under tightly controlled conditions of a clinical trial, but have not been used for long enough or widely enough to really understand their risks.

YELLOW

Elevated Risk
Create a personal risk reduction plan with your doctor
These are drugs that require a special discussion between you and your doctor about ways that together, you can minimize your drug risk. Such a plan could simply involve education about signs and symptoms to look out for, or could involve regular check-ups and monitoring.

ORANGE

High Risk
Requires careful consideration of risk versus benefit
These are drugs that are associated with a high risk with serious side effects, and require a special discussion between you and your doctor to discuss risk versus benefit.

RED

On Hold - Risk Rating under review

GREY

Source: iGuard [32].

the physical fitness side, games such as the Nintendo Wii, and in particular the Wii*Fit* [33] program, may prove particularly useful for the elderly, as they combine entertainment with physical movement. Osteoporosis and joint pain could be improved via muscle toning, all while having fun and interacting with others nearby or far away (such as family members). Motor

Table 7.3 Possible benefits of gaming for the elderly.

Physical	Mental	Emotional	Social
Hand–eye coordination	Attention span	Self-motivation	Decision-making
Fine motor skills	Short-term memory	Self-confidence	Social integration
Muscle toning	Problem solving	Self-esteem	Less isolation
Perceptual adaptability	Critical thinking	Emotional well-being	Collaborative thinking & team work

skills, eye–hand coordination and dexterity can be further trained, and provide heightened confidence to the user. Physical rehabilitation and geriatric wards could profit from gaming applications that are enhanced by virtual reality environments.

In addition to physical benefits, gaming can be helpful for the maintenance of cognitive abilities, such as memory and decision-making. Keeping the mind active becomes a greater challenge after retirement, particularly if social interaction is limited. General well-being is enhanced when individuals feel mentally agile, and better able to remember things.

In order to reap these benefits, however, it is imperative that technologists and designers create interfaces and communication systems that are geared for elderly people that may have visual and/or hearing impairments. Important considerations include screen design, operating procedures, (e.g., menus), instrumentation, and training mechanisms. Usability poses a much bigger barrier to adoption for older users than younger users.

Transacting

Another important element of web 2.0 applications and real-time information processing on-line is the streamlining of routine tasks, and the added convenience of ordering goods without having to leave the home. In a number of economies, the elderly, and busy young people, are already taking advantage of on-line grocery shopping. Many service providers offer delivery, and remember previous orders and preferences to make the task simpler. Ordering medical supplies and prescription medication can also be facilitated in the same way, and is already being done in some areas. Services such as requests for homecare visits or doctor's visits should be encouraged, as should other e-government applications such as easy access to pension and disability information.

7.3.2 System Cognition and Adaptability

The renowned French architect Le Corbusier once wrote that homes should be "machines to live in." A further step in the development of computing is the creation of machines capable of learning, thought and adaptation, making them responsive to constantly changing environments. The ability of human beings to think and adapt can primarily be seen through their ability to deconstruct their own thinking (i.e., meta-cognition) and to "muddle through" problems even when they seem to be unusual or without clear or predictable outcomes. If computer systems were to be able to "muddle through" complex and unpredictable problems in this manner, they would be better able to respond in a crisis or assist individuals that may require varying levels and degrees of help. Equipped with such intelligence, the home itself could become smart, adaptive and responsive to an inhabitant's changing needs and preferences. This cognitive ability is being increasingly explored in the context of robotics and many predict that home robots may soon become as common as dishwashers. Robots would interact not only with people, but also with various things they come in contact with. Their role will become increasingly important in fuelling context-awareness living.

In this context, the use of computer systems for enhancing human memory is an interesting development [34]. Designing technologies for memory amplification is a new area of research: this includes devices that interact with the environment and other objects in their vicinity in order to classify individual experiences in easy-to-retrieve formats for users. Past memories or facts relevant to a user's current context can be raised and accessed when appropriate. This is a particularly valuable development for the elderly and those suffering from loss of short-term memory and other cognitive problems.

Of course, there are a number of technical challenges to overcome, with some areas requiring additional heavy investments in advanced R&D. Quite apart from these challenges, other major hurdles to the development of adaptable and context-based thinking homes and environments, include data accuracy, data overload, data protection, and privacy.

7.3.3 Virtual Worlds for Ambient Living

Perhaps one of the most interesting platforms for the elderly is what might be called ambient co-living. Even before systems become cognitive, they can be ubiquitous, or embedded in the environment around us. For instance, imagine a large screen, placed like a painting on a wall, with animated avatars representing family members. These avatars could be animated, as desired, with real-life data, when individuals return home, for instance, indicating meal times, playtime, bedtimes and so on. In this way, family members far away can feel as though they are living alongside with their loved ones, and also keeping an eye on daily routines and well-being. If an elderly parent falls asleep in their chair, or has left the stove on, this information could be passed along in a real-time fashion to families and/or other caregivers. This virtual "togetherness" or co-presence can help fight feelings of isolation and vulnerability. However, fears concerning the decline of face-to-face communications and interaction must be addressed. Ambient co-living should not replace personal contact, however, and have the unintended consequence of trapping the elderly in the own homes.

7.3.4 Scenario 2030: A Day in the Life

As discussed above, perhaps one of the most important advances to change the future landscape will be human–computer interfaces (HCI). Much research and development is presently underway. Commercial applications will most likely surface over the next decade or so. In order to better envisage what the future may look like as a result of forthcoming advances, notably in wireless networks, sensor technologies and HCI, let us attempt to paint a picture of what life might be like for an 80-year old about twenty years from today, say in 2030.

Vision 2030: A day in the life of Marianne, 80

At 80, Marianne is in relatively good health, but suffers from arthritis, visual impairment, asthma, lack of balance and mild cognitive impairment (MCI). She lives alone in a one-bedroom flat, about 40 km from her only daughter, Ida, 45, and her two children. Marianne's flat is embedded with a smart wireless ambient living system, allowing her to continue to live at home, despite her various medical conditions.

At 7:00 am, Marianne is awoken by the smell of coffee from the kitchen and the sound of her grandchildren's voices as they get ready for school being

---

(Continued)

(*Continued*)

transmitted from her daughter's home. She gets up, goes to the display in her bedroom, and touches the screen to indicate to Ida that she is also awake. Ida can respond by either touching the screen and sending a message, or speaking to Marianne and reminding her to do her exercises. A virtual hug is passed between mother and daughter though their avatars on the screen.

Marianne sits down in the kitchen with her cup of coffee and considers her activities for the day. The fridge is stocked full of her favorite groceries, as these were automatically ordered and delivered the previous day. Suggested recipes maximizing the intake of fibre, iron, and calcium are displayed on the fridge's screen and can also be read out loud for Marianne, if she wishes. Alternatively, she can choose from a wide variety of pre-prepared meals that are delivered with her grocery shopping. Marianne takes important heart and blood pressure medicine, together with doses of cortisone, for the morning through an automatic dispenser. If she had forgotten, a message would have been sent to Ida to remind her.

After breakfast, Marianne does some brain training, posture and reflex exercises with her friendly service robot, Abo. She communicates with the robot through voice and wrist/finger controls monitor her movement. Results on problem-solving and memory are tabulated on a monthly basis to evaluate the rate at which Marianne' cognitive impairment is deteriorating, if at all. The sensor-based motion training is intended to sharpen motor functions, hand–eye coordination and flexibility, which can worsen with age. Sometimes, Marianne plays sometimes remote games with her grand-children and other relatives and friends who live outside her native Denmark.

Marianne then decides to take a walk by the sea. The watch she is wearing is embedded with location technology, heart rate and blood pressure sensors. It can warn Marianne if she is overdoing it, and can guide her home if she loses her way. It can also contact emergency services and/or her daughter Ida in case of an unusual reading of her vital signs.

After returning home and eating lunch, Marianne takes an afternoon nap, as she is very tired from her long walk. While she sleeps, Robot Abo does some laundry. Around 18:00, as Marianne is still sleeping, Robot Abo sends a message to Ida who has just returned home. Ida tells Abo to let Marianne

(*Continued*)

(*Continued*)

> sleep for another 45 minutes, but then to wake her up and make sure she is okay. Before 18:45, Robot Abo detects that Marianne is wheezing in her sleep (through the sound sensors in the room), and wakes her to give her the ventolin asthma inhaler. He informs Ida, and Marianne then reassures her daughter using Abo's voice programme.
>
> After eating a healthy pre-prepared dinner, Marianne says goodnight to her grandchildren and spends some time chatting with her daughter Ida before retiring to bed. Robot Abo reads to her, as she slowly drifts to sleep. All in all, it has been a good day.

7.4   Socio-ethical Considerations

7.4.1   Orwellian Fears

Emerging technologies and applications, such as RFID, sensors, the internet of things, and web 2.0, revolutionize transaction speed, and automate routine tasks, to the point of becoming invisible. Mass deployment of these technologies is likely to enhance the quality of life, e.g., make life more convenient and safe reduce hassles and waiting times for services; and assist in the care of children, the elderly and infirm. The development of ambient or ubiquitous networking environments will enable and encourage the practice of flexible working hours, thereby reducing commuting times and having a positive effect on the environment and on human relationships (e.g., family). But no technological development is possible without effects upon society, be they desirable or undesirable, and concerted efforts must be made to better understand the socio-ethical challenges faced by modern society in the new digital age.

As real world activities and everyday items become mapped in the virtual world, information becomes subject to transmission by multiple networks in multiple contexts. For example, the political climate in most, if not all areas of the world, leads to the widespread concern that a culture of mass human surveillance exists and is rapidly growing. This has important ramifications for quality of life, privacy, individual empowerment, and social development. Wireless identification technologies combined with the internet will make it easier for governments and industry to monitor activities and location of citizens, through information made available by RFID-type tags planted on persons, places or things.

The use of such personal information needs to be carefully monitored and limited in scope, through legislation and guidelines not unlike the spirit of the laws of robotics as described in *I, Robot* by the science-fiction author Isaac Asimov, i.e.,

- *A robot may not injure a human being or, through inaction, allow a human being to come to harm.*
- *A robot must obey orders given to it by human beings, except where such orders would conflict with the First Law.*
- *A robot must protect its own existence as long as such protection does not conflict with the First or Second Law.*

The mechanism, purpose, and extent of identification (be it RFID data, biometric data etc…) must favor the citizen and adhere to principles of individual choice and transparency. This will prevent the development of an Orwellian culture of surveillance. At the present stage, it must be noted that these important concerns are being swept aside in the helter-skelter race for development and innovation at any cost.

Moreover, as ICTs become increasingly ubiquitous, more details about individual citizens, such as habits, preferences and behavioral patterns, can be gathered and processed. Such data acquisition may well have its beginning in purpose-specific interests, such as manufacturing, inventory control, home automation, and health care. But with the growth of concerns over public security and safety, data thus acquired will almost certainly be turned to sinister uses.

Increasingly, various public authorities and commercial organizations require people to complete forms on-line, e.g., news alerts. In many instances, this mechanism is often welcomed by the public. However, such is the fear of surveillance that prevails in their minds that people regularly provide fictitious information. It is obviously easier to do this remotely, than when they are in a face-to-face situation. This phenomenon makes data gathered for legitimate purposes unreliable. But today, people still retain a good amount a control over their data in that some avoid using the internet either deliberately or through ignorance of the services available. But with the coming internet of things, a wealth of data about the owners and users of "things" will be available, much of it involuntary, therefore beyond their control.

On the other hand, surveillance devices are growing in number and sophistication. Reports of surveillance devices under development are not entirely science fiction. Doorways that can read fingerprints are here today, and insect-like devices that float in and out of houses and buildings may not be as far away as we think. With reports of developments such as these, fears of surveillance are bound to grow, and exacerbate the present climate of distrust. This climate, whether real or perceived can foster anxiety in the most minor decision-making. Since the exercise of choice and the making of decisions is crucial to individual self-fulfilment and self-expression, it is also essential to the advancement of society. Paranoia in this regard will generally inhibit creativity and the maintenance healthy social intercourse.

Although technologies like RFID will reduce the number of routine tasks in daily life, the complete automation of human activity may not necessarily be a desirable outcome. It may lead to a sterile and uniform society. The need to conform may become greater, which in turn will hamper self-expression and individuality. Any deviation from standard practice would be recorded and viewed unfavorably. The OECD describes this phenomenon as "social sorting" (OECD, 2004). This is now at the basis of an entire industry devoted to clustering different populations. The use of RFID to mark and label all individuals should be discouraged, for it will eventually lead to a form of human "sorting" and "classifying," even if not actually first introduced for that purpose.

The considerations described above are particularly important for the elderly, as they are a relatively vulnerable group in society. They are, for instance, more likely to divulge personal information when requested, than their younger counterparts. They are more likely subjects of fraud and cyber-crime, because they are less familiar with current methods for illegally extracting information from individuals. They are also generally a more trusting group, and in some sense, less sceptical about seemingly *bona fide* commercial or government entities seeking to collect information. Therefore, special attention must be paid to this group in developing mechanisms for privacy and data protection, especially in light of developments in tracking technologies for health care and independent living.

7.4.2 Social Interactions in the Digital Space

Just as the use of the mobile phone (and in particular SMS), now mediates more and of more of human relationship [35], RFID might be yet another technology to threaten human intimacy. Human relationships seem increasingly transient and ephemeral [36] in an always-on digital age. Although technical devices, tagged household items and so on, may eventually become invisible to us, human-to-human communication and interaction should remain squarely in the limelight.

The rise of technologies that permit continuous or uninterrupted communications seems to further exemplify the age-old human struggle to conquer space and time. Our need to be connected with the past and the distant is the reason we record history, the reason we travel, the reason we look to the stars.

In some domains, in the efforts to conquer time, it has at least been marginalized. More and more, the initiating individual's convenience dictates the when and how of the communication that takes place ("communication on my own time"). There are advantages. Communication can be planned and managed more conveniently and efficiently. But communication is a two-way street. Such tailored communication "on my own time" can mean selective response and such techniques as resorting to voice mail or voice messaging. This may frustrate the maintenance of communications and actually lead to a downturn in the felicity of life and living.

Not only can we, as individuals, play a little with space and time through the use of technology, so too can commercial interests. For instance, spam and unsolicited marketing efforts can be inflicted upon us at any time and any place, potentially filling up every nook and cranny of our daily lives.

Another issue that is raised relates to the notion of intimacy and the ambiguous nature of communication. Intimacy is here understood as the feeling of closeness with another, to the exclusion of everyone else. Jane Austen puts it thus: "It is not time or opportunity that is to determine intimacy — it is disposition alone. Seven years would be insufficient to make some people acquainted with each other, and seven days are more than enough for others" [37].

The constant availability of communications and information has affected the degree of intimacy in relationships and the overall nature of social interaction. A good example is the preponderance in digital environments of "asynchronous" and "pre-rehearsed" communications. This means that the expression of thought or feeling is more often contrived, while truer gut reactions and spontaneous expressions become limited. As a result, an important set of clues or signs that individuals require to build relationships and establish trust are lacking. One cannot expect human beings to make choices and decisions about their social relations based solely on statistics and pre-prepared dialogue.

Another interesting aspect is the rise of digital gesturing: from the "how r u" SMS to the IM (instant messaging) "poke" or "nudge." There is much ambiguity in it. The IM poke, for instance, is merely a sign that can take the form of a shaking or a noise on the recipient's side. This signifies a desire to be in touch, but no more. Whether the communication is postponed, desired, or simply not required, is unclear, perhaps even to the sender. It leaves the communication open-ended and with uncertain implications.

Also in IM systems, users can remain in "invisible" mode, often known as "stealth" mode. In this manner, users can observe, unseen, other users when they come on-line and off-line. They are unreachable but are able to contact available/visible users who come on-line. This is another example of "communication on my own time, and on my own terms." An opposite phenomenon is "always-on exhibitionism," when users wish to be seen on-line at all times and provide a maximum amount of detail about their person (either true or falsified information). In social networks, such users also serve "communication hubs" or brokers for creating additional connections to the network. Not surprisingly, showing off friends and connections in public is a way to impress and to demonstrate social status, both in the on-line and off-line worlds.

As there is a lack of nuance, however, in these on-line social spaces, it is difficult to gain a good understanding of another user's personality. Many can be falsified, in an obvious way (like "fakesters" who use favorite bands or actors as their username) or in a less obvious way, when users hide those aspects of their personality that they consider unsuccessful in the physical world. Personalities on-line are also subject to constant change, as people continuously re-invent themselves. The projection of identity, and the re-writing of it on-line, is an important preoccupation of today's digital world.

The new technologies that are in the course of rapid development have been accompanied by the birth and growth of a so-called sub-culture of communications, with its own identity formulation, etiquette and social norms in constant flux. This sub-culture impacts many aspects of life and needs to be mapped and studied carefully. Naturally, most aspects of matters such as this are not subject to a formalized protocol. Nevertheless, ways must be found to foster thinking and education about norms and standards of technology use and behavior in the public interest.

It must be noted that in this context for the elderly, in particular, it is even more difficult to conform and adjust to new developments in digital social norms. Not only do they have to unlearn years of behavior, but by virtue of their advanced age, learning new ways of doing things can become a great burden. However, it also presents an opportunity, as independent living technologies for instance could significantly enhance quality of life and convenience. What is required is a greater investment in education not only about the technical aspects of new applications but their social ramifications as well.

In any human enterprise, not only should one seek to maximize profit, but we must also reduce or eliminate loss. In Dr Faustus, that mythical bargain was eventually turned in our favor. One can hope that the Faustian Bargain of technological change [38] will also follow that pattern: that on balance, the gains that new technologies generate will more than compensate for whatever downsides they may have [39]. But for this to be possible, an awareness of the subtle impacts of technology on human behavior and society is necessary at the earliest stages of innovation. Only an understanding of the ultimate limitations of new technology will enable us to fully reap its seemingly limitless potential.

7.4.3 Age, Civilization, and the Machine

Self-preservation has long been recognized as the most basic of instincts. An extension of this is the instinct to preserve and promote society. In turn, society, or the presence of others, is necessary for the preservation of the individual.

Civilization is in part the result of the cumulation of human experience in society. This encompasses social norms, social structures and hierarchies, and also, more fundamentally, human thought and contemplation. In this sense, aged humanity is our repository of experience, and thus the custodian of culture and civilization. If the advance of civilization is indeed our most desirable aim, then the aged stage in life, a state to which we all may aspire, is eminently worth preserving and empowering. And this requires mechanisms of societal support.

Life, and particularly human life, depends on support from its very beginnings: financial, physical, medical, social and such. The newly born require intensive nursing: their very survival depends on it. Children require support to learn. In adulthood, people require domestic help, childcare support and technical assistance with household maintenance and increasingly complex home appliances. It is a commonly observed fact that with the onset of age, the need for support, particularly physical and medical support, grows. At the same time, the number of younger persons who might potentially provide that support is shrinking as the demands on their time and energy increase. As it happens, this gap could be bridged by emerging technologies such as those described in this book. It may be worth recalling what Oscar Wilde wrote over a hundred years ago: "unless there are slaves to do the ugly, horrible, uninteresting work, culture and contemplation become almost impossible … on mechanical slavery, on the slavery of the machine, the future of the world depends" [40]. Paraphrased for modern times, it can be said that civilization has always required support to flourish — in the future, this support will come increasingly from "machines," which are growing not only in quantity but also complexity and diversity, and perhaps most importantly, in their ability to interact with, and adapt to, the human context.

8

Conclusions and Future Directions

Beautiful young people are accidents of nature, but beautiful old people are works of art.

— Eleanor Roosevelt

8.1 Summary

This book has presented a proposal for a home wireless platform for the elderly, in view of changing demographic trends, widening public policy objectives, and continuing developments in emerging technologies. It has proposed the integration of traditionally disparate technologies, in order to present what might be described as an early prototype of a digital home. In the course of this effort, a new decision-support system for behavior monitoring in the home has been developed. Computer simulation has been used to verify the system. Full account has been taken of socio-ethical and multi-disciplinary perspectives and the need for these to play a greater role in innovation and technological design.

Chapter 2 presented a forward-looking overview of the social influences shaping technological adoption and how society in turn affects technological development. It drew attention to the changing demographics around the world and the problems faced by ageing societies. It highlighted growing public policy involvement in stimulating the adoption of emerging information and communication technologies as well as its important social priorities (such as health care and ageing). Given the synergies between these policy priorities, it was plain to see that the time is now right for applying emerging ICTs to enhancing the quality of life for the aged.

Chapter 3 cast new light on user-centric design and demand-pull strategies. It proposed a new theoretical framework for "use leadership" in innovation, based on von Hippel's original notion of lead users, and applied that framework to the case of elderly and emerging technologies.

Chapter 4 discussed a key emerging technology, RFID — its technical aspects, markets and unique potential. It explored its many applications, advantages, and functionalities before addressing its limitations. The chapter examined the role of RFID in health care and summarized the main gaps. One of the important assertions made was how RFID can best complement, and be complemented, by other technologies, such as sensors and wireless networking. It concluded by exploring the widening context for RFID and its future prospects.

Chapter 5 began by setting out the main challenges in networking the home, before presenting the background and objectives of the home system for elderly care, known as the AGE@HOME Platform. It described the overall capabilities of the system and the basics of the behavior monitoring model required for crisis alerts. It also introduced the methodology used for the logic diagrams that followed in Chapter 6.

Chapter 6 presented logic diagrams based on SDL (Specification and Description Language) to describe the AGE@HOME decision-support system. This decision logic was used to develop the AGE@HOME software program in MATLAB. Chapter 6 described this program and its various components, such as sensor checks, timer controls, control interfaces, and object localization. It explained how administrators could set up parameters to suit specific home environments or resident needs. It also presented the results of the random simulation, which confirmed the decision logic as initially conceived, and verified that the system could be applied to a real-life scenario, in which physical sensors provide real data. The software was observed to perform the role for which it was intended, i.e. a behavioral and environmental monitoring system. Finally, this chapter raised some of the important challenges for technology developers, users and policy-makers.

Chapter 7 explored the new frontiers of digital living. It examined the evolution of daily life in the digital age, taking into account developments such as the participatory web and the network of things, including implications for the elderly. It highlighted the need for user-centred innovation, and the creation of an all-inclusive and personalized "web i.o." Finally, it explored some of the socio-ethical perspectives needed for more effective and human technology design.

The AGE@HOME concept and platform take into account technological systems available today, but can be upgraded with new technology as it becomes available. Such modifications are made simpler because the system itself focuses primarily on the needs of users, and, in a wider sense, addresses socio-economic challenges that will endure for some time.

8.2 From Theory to Practice

The theoretical concept presented here is ready for laboratory testing. A laboratory could utilize the computer program and user interface that has been developed and put into practice the physical integration of the various technical components. In particular, the laboratory could test how RFID tags and sensors can be best combined on one device, and explore the networking capabilities of readers and wireless LAN. The network configuration could be further tested, e.g., by taking into account anti-collision mechanisms and reader ranges. The laboratory could further test the computer program with actual real-time sensor and tag data. The unobtrusive nature of the wireless system should also be ensured through the use of smaller and well-placed components.

The laboratory could be set up in a research environment, or alternatively, as a "living laboratory" deployed in different elderly homes for testing. It could also be tested in institutional settings (e.g., nursing homes). In order to truly put theory into practice, however, surveys and discussions groups should be used to gauge user acceptance (including technical awareness and ease of use). These should include the elderly but also their main caregivers, personal and professional. Finally, further work could be conducted on the user interface through targeted user feedback and usage monitoring.

8.3 Opportunities for Further Work

The platform as it is designed has the potential to be expanded further, through the addition of new components, in light of burgeoning technology and tailored individual need. For instance, other types of sensors could be added, such as biomedical sensors ranging from glucose meters to blood pressure sensors. Similarly, the field of application could be expanded to cover residents with a wider range of disabilities. The addition of accelerometers and more sophisticated motion sensors could help with the determination of the precise location and motion patterns of a resident within the home. This would enable the collection of more accurate health and behavioral data in the home environment, which could eventually be shared with professional caregivers (if necessary and appropriate). It may also be of use in institutions, such as care homes for the elderly or psychiatric institutions. It will be vital in all cases, however, for further work to be carried out on the security and privacy parameters of the data collected by such a system. Standardization of the technology in use is also an important challenge that needs further concerted effort at the global scale.

Furthermore, the system could also be used to determine the level of energy and water consumption in a household. This has potential benefits not only for elderly households, but for a much wider user base, given the increased need felt globally for resource and energy efficiency. Like ageing, energy is an important public policy priority. Future systems based on the proposed platform could integrate actuators with sensors, for the real-time regulation and adjustment of consumption, or for addressing unusual environmental conditions, such as excessive water flow or abnormal ambient temperature. The behavior monitoring model proposed here might be extended to other probability-based applications, and be of use for diverse purposes, such as patient flow and behavior in hospitals, visitor flow in public areas, or building maintenance.

Further work is required on the web-based user interface, in particular the addition of voice and speech recognition tools, which may be of particular assistance to elderly or handicapped that have trouble with the physical input of commands. This interface could also be further extended to reap the advantages of social networking systems, thus enabling lifelong learning and community living. Gaming, and in particular "lifestyle gaming," should be considered an important development for the elderly. Daily chores and health maintenance can be made entertaining and thus ease the burden on those who may have difficulty in completing them. This can then be extended to other segments of the population.

Finally, the proposed platform may eventually lead to ambient systems for co-presence, whereby caregivers might observe, in a digital space (e.g., through animation and avatars), the real-time behavior of the elderly in their care. This would provide greater continuity of care and facilitate the "seamless" incorporation of elder care into daily life. In a larger sense, it would help further bridge the gap between geographical spaces (here and there) and between real and virtual spaces, thereby enhancing and further shaping the scale, scope, and impact of today's digital networks.

References

Chapter 1

[1] European Commission, "Eurostat pocketbook." *Living Conditions in Europe 2002–2005*, 2007.

Chapter 2

[1] L. Srivastava, "The regulatory environment for future mobile multimedia services," *Computer and Communications Law Review*, nos. 7 & 8, 2006.

[2] CORDIS, Seventh Framework Programme, Keywords, available at http://cordis. europa.eu/fp7/ict/browse/keywords_en.html#c (accessed 7/7/2008).

[3] Facebook, Home Page, available at http://www.facebook.com/.

[4] OECD database, Ageing and Employment Policies and UN, World Population Prospects 1950–2050 (The 2002 Revision) for the Russian Federation.

[5] European Commission, Communication from the Commission: Green Paper "Confronting demographic change: A new solidarity between the generations," COM(2005) 94 Final, March 2005.

[6] OECD, "Ageing populations: High time for action," Background paper prepared by the OECD Secretariat, Meeting of G8 Employment and Labour Ministers, London, 10–11 March 2005.

[7] BBC News, "India's demographic dividend," 25 July 2007, available at http://news.bbc.co.uk/2/hi/south_asia/6911544.stm (accessed 31/07/2008).

[8] C. Haub, "Global aging and the demographic divide," *Public Policy and Aging Report*, vol. 17, no. 4, 2007.

[9] European Commission, Eurostat Statistical Books, The life of women and men in Europe: A statistical Portrait, 2008. This figure does not include Denmark, Ireland or Sweden, and excludes those living in nursing home care.

[10] OECD, *OECD Factbook 2008: Economic, Environmental and Social Statistics*, 2008.

[11] E. Von Hippel, *Democratizing Innovation*. Cambridge MA: MIT Press, 2005.

[12] E. Von Hippel, "Lead users: A source of novel product concepts," *Management Science*, vol. 32, no. 7, 1986.

[13] E. von Hippel, *The Sources of Innovation*. New York, NY: Oxford University Press, 1988.

[14] L. Srivastava and R. Mansell, "Electronic cash and the innovation process: A user paradigm," SPRU Electronic Working Paper no. 23, 1998.

Chapter 3

[1]   E. Von Hippel, *Democratizing Innovation*, Cambridge, MA: MIT Press, 2005.

[2]   E. M. Rogers, *Diffusion of Innovations*, fourth ed., New York, NY: The Free Press, 1995.

[3]   J-J. Salomon, F. R. Sapasti, and C. Sachs-Jeantet (eds.), *The Uncertain Quest: Science, Technology and Development*, United Nations University Press, 1994.

[4]   E. Von Hippel, "Lead users: A source of novel product concepts," *Management Science*, vol. 32, no. 7, 1986.

[5]   M. Schreier and R. Prugl, "Extending lead user theory: Antecedents and consequences of consumers' lead userness," *Journal of Product Innovation Management*, vol. 25, no. 4, 2007.

[6]   E. Von Hippel, "Lead users: A source of novel product concepts," *Management Science*, vol. 32, no. 7, 1986.

[7]   G. Urban and E. Von Hippel, "Lead user analyses for the development of new industrial products," *Management Science*, vol. 34, no. 5, 1988.

[8]   D. W. Rae and M. Taylor, *The Analysis of Political Cleavages*, New Haven, CT: Yale University Press, 1970.

[9]   P. D. Morrison, J. H. Roberts, and D. F. Midgley, "The nature of lead users and measurement of leading edge status," *Research Policy*, vol. 33, 2004.

[10] Euromonitor International, "The pensioner market: Old age purchasers," *Report on The Swing Generation: Marketing to the Over 65s*, Euromonitor, March 2007.

[11] C. de Koninck, *La Primauté Du Bien Commun Contre Les Personnalistes*, Éditions de l'Université Laval, Montréal, Éditions Fides, 1943.

Chapter 4

[1]   H. Stockman, "Communication by means of reflected power," *Proceedings of the IRE*, October 1948.

[2]   G. E. Moore, "Cramming more components onto integrated circuits," *Electronics*, 19 April 1965.

[3]   J. Landt, "Shrouds of time: The history of RFID," *AIM*, 2001.

[4]   A. Koelle, S. Depp, and R. Freyman, "Short-range radio-telemetry for electronic identification using modulated backscatter," *Proceedings of the IEEE*, vol. 63, no. 8, August 1975, available at http://ieeexplore.ieee.org/iel5/5/31195/01451858.pdf (accessed 25/03/08).

[5]   RFID Journal, "Wal-Mart draws line in the sand," 11 June 2003, available at http://www.rfidjournal.com/article/view/462/1/1/ (accessed 25/03/2008).

[6]   RFID Journal, "Wal-Mart expands RFID Mandate," 18 August 2003, available at http://www.rfidjournal.com/article/articleview/539/1/1/ (accessed 25/03/08).

[7]   M. H. Weier, "Wal-Mart gets tough on RFID," *Information Week*, 19 January 2008, available at http://www.informationweek.com/news/showArticle.jhtml;jsessionid=EUWND-S2DXA1SCQSNDLOSKHSCJUNN2JVN?articleID=205900561 (accessed 25/03/08).

[8]   M. L. Katz and C. Shapiro, "Network externalities, competition, and compatibility," *American Economic Review*, vol. 75, no. 3, 1985.

[9]   L. Srivastava, "Radio frequency identification: Ubiquity for humanity," *INFO*, vol. 9, no. 1, 2007.

[10] BBC News, "World's tiniest tag unveiled," 23 February 2007, available at http://news.bbc.co.uk/2/hi/technology/6389581.stm (accessed 28/03/08).

[11] R. Want, "RFID: A key to automating everything," *Scientific American*, January 2004.

[12] Auto-ID Center, White Paper, "Active and passive RFID: Two distinct, but complementary, technologies for real-time supply chain visibility," available at http://www.autoid.org/2002_Documents/sc31_wg4/docs_501-520/520_18000-7_WhitePaper.pdf (accessed 31/07/2008).

[13] EPC, Home Page, available at http://www.epcglobalinc.org/home (accessed 31/03/08).

[14] S. Meloan, Sun Developer Network, "Towards a global internet of things," 2003, available at http://java.sun.com/developer/technicalArticles/Ecommerce/rfid/ (accessed 31/07/2008).

[15] UID Center, Home Page, available at http://www.uidcenter.org/index-en.html (accessed 31/03/08).

[16] International Organization for Standardization, Home Page, available at http://www.iso.org/iso/home.htm (accessed 31/03/08).

[17] M. Weinländer, "What is EPC?," *RFID systems SIMATIC RF*, Siemens, 2006, available at http://www.automation.siemens.com/download/internet/cache/3/1455039/pub/de/wp_rfid_epc_e.pdf (accessed 31/07/2008).

[18] K. Finkenzeller, *RFID Handbook — Fundamentals and Applications in Contactless Smart Cards and Identification*. Second ed., New York, NY: Wiley, 2003.

[19] K. Sakamura, "Ubiquitous ID technologies 2008," YRP Ubiquitous Networking Laboratory, uID Center, 2008, available at http://www.uidcenter.org/pdf/UID910-W001-080226_en.pdf (accessed 31/07/08).

[20] IDTechEx, "RFID forecasts, players & opportunities 2008–2018," at http://www.idtechex.com/products/en/view.asp?productcategoryid=151 (accessed 1/04/08).

[21] Gartner, Press Release, "Gartner says worldwide RFID revenue to surpass $1.2 Billion in 2008," at http://www.gartner.com/it/page.jsp?id=610807 (accessed 1/04/08).

[22] Centredoc, "RFID intellectual property database," available at http://www.centredoc.ch/en/homeE.asp (accessed 31/03/08).

[23] J. Fenn, "Understanding Gartner's Hype cycles, 2007," 5 July 2007, available at http://www.gartner.com/DisplayDocument?id=509085&ref=g_SiteLink (accessed 31/03/08). (See also RFID Weblog at http://www.rfid-weblog.com/archives/market_size.php (accessed 31/03/08)).

[24] G. Nanda, V. M. Bove, and A. Cable, "bYOB Build your own bag: A computationally-enhanced modular textile system," available at http://alumni.media.mit.edu/~nanda/design/electronics/byob/papers/bYOB_UbicompDemos.pdf (accessed 2/04/08).

[25] ITU (ed. L. Srivastava), ITU Internet Reports 2005: Internet of Things, Geneva, 2005, available at http://www.itu.int/internetofthings/ (accessed 20/08/08).

[26] K. Sakamura, "The world of computers everywhere: Ubiquitous computing and its application," Presented at the Royal Swedish Academy of Engineering Sciences, November 2007, available at http://www.iva.se/upload/Verksamhet/Projekt/ Internet-framsyn/U%20Japan%20Sakamura%20071122.pdf (accessed 14/04/08).

[27] X. Liu, M. D. Corner, and P. Shenoy, "Ferret: RFID localization for pervasive multimedia," Department of Computer Science, University of Massachusetts, in *Proceedings of the 8th Ubicomp Conference*, September 2006, available at http://prisms.cs.umass.edu/mcorner/papers/ubicomp_2006_ferret.pdf (accessed 4/04/09).

[28] J. Hightower, R. Want, and G. Borriello. "SpotON: An indoor 3d location sensing technology based on RF signal strength," UW-CSE 00-02-02, University of Washington, Department of Computer Science and Engineering, Seattle, WA, February 2000, available at http://seattle.intel-research.net/people/jhightower/pubs/hightower2000indoor/hightower2000indoor.pdf (accessed 3/04/2008).

[29] J. Hightower, C. Vakili, G. Borriello, and R. Want, "Design and calibration of the spotON ad-hoc location sensing system," August 2001, unpublished, available at http://seattle.intel-research.net/people/jhightower/pubs/hightower2001design/hightower2001design.pdf (accessed 3/04/08).

[30] L. M. Ni, R. Liu, Y. Cho Lau, and A. P. Patil, "LANDMARC: Indoor location sensing using active RFID," *Wireless Networks*, vol. 10, 2004, available at http://www.cs.ust.hk/~liu/Landmarc.pdf (accessed 31/07/08).

[31] D. Hahnel, W. Burgard, D. Fox, K. Fishkin, and M. Philipose, "Mapping and localization with RFID technology," University of Freiburg, University of Washington, Intel Research Seattle, available at http://seattleweb.intel-research.net/people/matthai/pubs/icra04.pdf (accessed 4/04/08).

[32] G. Borriello, W. Brunette, M. Hall, C. Hartung, and C. Tangney, "Reminding about tagged objects using passive RFIDs," available at http://www.uwnews.org/related-content/2004/October/rc_parentID5748_thisID5749.pdf (accessed 4/04/08).

[33] V. D. Hunt, A. Puglia, and M. Puglia, *RFID — A guide to Radio Frequency Identification*. Wiley & Sons, 2007.

[34] The Register, "Feds approve human RFID implants," 14 October 2004, available at http://www.theregister.co.uk/2004/10/14/human_rfid_implants/ (accessed 4/04/08).

[35] A. Solanas and J. Castella-Roca, "RFID technology for the health care sector," *Recent Patents on Electrical Engineering*, vol. 1, 2008, pp. 22–31.

[36] P. Nagy, I. George, W. Bernstein, J. Caban, R. Klein, R. Mezrich, and A. Park, "Radio frequency identification systems technology in the surgical setting," *Surgical Innovation 2006*, vol. 13, no. 1, March 2006, pp. 61–67.

[37] BRIDGE Project, Logica CMG and GS1, "European passive RFID market sizing 2007–2022," February 2007, available at http://www.bridge-project.eu/data/File/BRIDGE%20WP13%20European%20passive%20RFID%20Market%20Sizing%202007-2022.pdf (accessed 31/07/08).

[38] European Commission JRC and Institute for Prospective Technological Studies (IPTS), "RFID technologies: Emerging issues, challenges and policy options," EUR 22770 EN, 2007.

[39] A. Coronato, G. D. Vecchia, and G. De Pietro, "An RFID-based access and location service for pervasive grids," *Emerging Directions in Embedded and Ubiquitous Computing (Lecture Notes in Computer Science)*, Springer (Berlin/Heidelberg), 2006, pp. 601–608.

[40] RFID Journal, "RFID sees gains in health care," 15 February 2006, available at http://www.rfidjournal.com/article/articleprint/2154/-1/1/ (accessed 11/04/08).

[41] IDTechEx, *RFID in Healthcare 2006–2016*, June 2006.

[42] Stanford School of Medicine, Press Release, "RFID chips can help surgeons avoid leaving sponges inside patients, study finds," 18 July 2006, available at http://healthpolicy.stanford.edu/news/rfid_chips_can_help_surgeons_avoid_leaving_sponges_inside_patients_study_finds_20060718/ (accessed 10/04/2008).

[43] RFID Journal, "RFID chips could aid surgical litter-bugs," 19 July 2006, available at http://www.theregister.co.uk/2006/07/19/rfid_surgery_debris/ (accessed 10/04/2008).

[44] A. Macario, D. Morris, and S. Morris, "Initial clinical evaluation of a handheld device for detecting retained surgical gauze sponges using radiofrequency identification technology," *Archives of Surgery*, vol. 141, no. 7, July 2006, pp. 656–662.

[45] P. Fuhrer and D. Guinard, *Building a Smart Hospital using RFID Technologies*. University of Fribourg, Switzerland.

[46] RFIDNews.org, "VeriChip and New Jersey health insurer begin 2-year trial of implantable RFID," 18 July 2006, available at http://www.rfidnews.org/weblog/ 2006/07/18/verichip-nj-health-insurer-begin-2year-trial-of-implantable-rfid/ (accessed 11/04/2008).

[47] R. Carlson, S. R. Silverman, and Z. Mejia, "Development of an implantable glucose sensor," Verichip White Paper, available at http://www.verichipcorp.com/files/ GLUwhiteFINAL.pdf (accessed 10/04/2008).

[48] Wikipedia, Definition of Sensor, available at http://en.wikipedia.org/wiki/Sensors (accessed 09/07/2008).

[49] L. Srivastava, "The mobile makes its mark," in *Handbook of Mobile Communication Studies*. (J. Katz, ed.), MIT Press, 2008.

[50] European Commission, "The disappearing computer proactive initiative in FP5 (1998– 2002)," available at http://www.cordis.lu/ist/fet/home.html.

[51] S. Roundy1, D. Steingart, L. Frechette, P. Wright, and Jan Rabaey, "Power sources for wireless sensor networks," *Book Series Lecture Notes in Computer Science*, Berlin: Springer, 2004.

[52] Wireless World Research Forum (WWRF) — Special Interest Group (SIG) 3, "Self-organisation in future mobile communication networks," April 2008.

[53] The Active RFID reader and Wi-Fi Hub provided by TagSense, at http://www. tagsense.com/ingles/products/product_mw.html (accessed 17/07/08).

[54] Zigbee Alliance, Home Page, available at http://www.zigbee.org/en/index.asp (accessed 17/072008).

[55] Bluetooth Special Interest Group, Home Page, available at https://www.bluetooth. org/apps/content/ (accessed 17/07/2008).

[56] Infrared Data Assocation, Home Page, available at http://www.irda.org/ (accessed 17/07/08).

[57] R. Prasad and K. Skouby, "Personal network (PN) applications," in *Wireless Personal Communications*, vol. 33, pp. 227–242.

[58] MAGNET website, available at http://www.ist-magnet.org/technicalapproach (accessed 31/07/2008).

[59] WIPO Patent, WO/2007/073473, "Acoustic wave device used as RFID and as sensor," available at http://www.wipo.int/pctdb/en/wo.jsp?IA=WO2007073473& wo=2007073473&DISPLAY=DESC (accessed 16/04/08).

[60] European Commission JRC and Institute for Prospective Technological Studies (IPTS), "RFID technologies: Emerging issues, challenges and policy options," EUR 22770 EN, 2007.

[61] RFID Journal, "R&D firm developing passive ultra-wideband RFID," 8 April 2008, available at http://www.rfidjournal.com/article/articleview/4010/ (accessed 16/04/08).

[62] RFID Journal, "Hospital gets ultra-wideband RFID," 29 August 2004, available at http://www.rfidjournal.com/article/view/1088/1/1 (accessed 16/04/08).

[63] M. Weiser , "The computer of the 21st century," *Scientific American*, vol. 265, 1991, pp. 66–75.

[64] A. Huxley, *Brave New World*, 1932.

Chapter 5

[1] BT, BT Home Monitoring Website, available at http://www.productsandservices.bt.com/consumerProducts/displayProduct.do?productId=CON-3295 (accessed 21/07/2008).

[2] Manodo, Manodo Home Page, available at http://www.manodo.se/site/295/hem.aspx (accessed 21/07/2008).

[3] Manodo, Product Folder — Manodo Home Arena, available at http://www.manodo.se/site/523/Default.aspx (accessed 31/07/08).

[4] T. Yamazaki, "The ubiquitous home," National Institute of Information and Communications Technology (NICT), Japan.

[5] S. Dixit and R. Prasad (eds.), *Technologies for Home Networking*. New Jersey, NJ: Wiley & Sons, 2008.

[6] Identec, Home Page, available at http://identecsolutions.com/ (accessed 14/07/2008).

[7] L. Srivastava, "The regulatory environment for future mobile multimedia services," *Computer and Communications Law Review*, nos. 7 & 8, 2006.

Chapter 6

[1] ITU, ITU-T, "Specification and description language (SDL)," ITU-T Recommendation Z.100 (Previously CCITT Recommendation), Series Z: Languages and general software aspects for telecommunication systems: Formal description techniques (FDT) — Specification and, Description Language (SDL), available at http://www.itu.int/ITU-T/studygroups/com10/languages/Z.100_1199.pdf (accessed 06/08/08).

Chapter 7

[1] L. Srivastava, "Japan's ubiquitous mobile information society," *INFO*, vol. 6, no. 4, 2004.

[2] Statistics Bureau of Japan, *Statistical Handbook of Japan 2007*. 2007.

[3] Japan Cabinet Office, 2008 White paper on the aging society, 2008.

[4] I. Fuyuno, "Ageing society in Japan," British Embassy, Tokyo, 2007, available at http://www.bcageing.org.uk/downloads/ageing%20society%20report%20part%20I.pdf (accessed 31/08/08).

[5] Ministry of Internal Affairs and Communications, white paper on information and communication in Japan, 2006.

[6] Ministry of Internal Affairs and Communications, Japan, e-Japan Priority Policy Program, available at http://www.kantei.go.jp/foreign/it/network/priority-all/index.html (accessed 01/08/08).

[7] Ministry of Internal Affairs and Communications, Japan, e-Japan II, available at http://www.kantei.go.jp/foreign/policy/it/0702senryaku_e.pdf (accessed 01/08/08).

[8] Ministry of Internal Affairs and Communications, u-Japan Policy, available at http://www.soumu.go.jp/menu_02/ict/u-japan_en/index.html (accessed 31/07/08).

[9] Ministry of Internal Affairs and Communications, The National ICT Strategies in Japan are evolving from "e" (electronics) to "u" (ubiquitous), 2002, available at http://www.soumu.go.jp/menu_02/ict/u-japan_en/new_outline01b_f.html (accessed 31/07/08).

[10] Ministry of Internal Affairs and Communications, u-Japan Policy, Policy package 2, available at http://www.soumu.go.jp/menu_02/ict/u-japan_en/new_pckg02_menu.html (accessed 31/07/08).

[11] Ministry of Internal Affairs and Communications, Presentation by A. Umino (Deputy Director), 2007, available at http://www.soumu.go.jp/joho_tsusin/eng/ presentation/pdf/071122_1.pdf (accessed 31/07/08).

[12] A. Gothenberg, "Japan's IT strategy for 2010 — a ubiquitous network society," *Growth Policy Current Affairs*, no. 2, May 2007.

[13] Silicon.com, "Schoolchildren to be RFID-chipped," 8 July 2004, available at http://networks.silicon.com/lans/0,39024663,39122042,00.htm (accessed 2/08/08).

[14] Ubiks, RFID in Japan, available at http://ubiks.net/local/blog/jmt/archives3/003065.html (accessed 02/08/08).

[15] The Guardian Online, "Tagging Tokyo's streets with no name," 10 May 2007, available at http://www.guardian.co.uk/technology/2007/may/10/japan.guardianweeklytechnology-section (accessed 31/07/08).

[16] L. Srivastava, "Ubiquitous network societies: The case of Japan," ITU New Initiatives Programme, 2005, available at www.itu.int/ni/.

[17] Ministry of Internal Affairs and Communications & Ministry of Economy, Trade and Industry (Japan), "Guidelines for privacy protection with regard to RFID tags," July 2004, available at http://www.meti.go.jp/english/information/data/IT-policy/pdf/guidelines_for_privacy_protection_with_regard_to_rfid_tags.pdf.

[18] TechRadar, "RFID keys for high-tech house of the future," 26 March 2008, available at http://www.techradar.com/news/phone-and-communications/mobile-phones/rfid-keys-for-high-tech-house-of-the-future-273395 (accessed 3/08/08).

[19] Reuters, "Japan makes robot girlfriend for lonely men," June 17 2008, available at http://www.reuters.com/article/technologyNews/idUST8462420080617 (accessed 06/08/08).

[20] Ministry of Internal Affairs and Communications, 2007 White Paper on Information and Communications in Japan, 2007, available at http://www.johotsusintokei. soumu.go.jp/whitepaper/eng/WP2007/2007-index.html (accessed 03/08/08).

[21] ITU (ed. L. Srivastava), ITU Internet Reports 2005: Internet of Things, Geneva, 2005, available at http://www.itu.int/internetofthings/ (accessed 20/08/08).

[22] ITU (ed. L. Srivastava), ITU Internet Reports 2006: Digital.life, Geneva, 2006, available at http://www.itu.int/osg/spu/publications/digitalife/ (accessed 20/08/08).

[23] The ITU-T Global Standards Initiative (GSI) on Identity Management (IdM), at http://www.itu.int/ITU-T/gsi/idm/ (accessed 21/07/08).

[24] T. O'Reilly, "What is Web 2.0: Design patterns and business models for the next generation of software," September 2005, available at http://www.oreillynet. com/pub/a/oreilly/tim/news/2005/09/30/what-is-web-20.html (accessed 21/04/08).

[25]   Britannica Online, Home Page, available at http://www.britannica.com/ and Wikipedia, Home Page, available at http://www.wikipedia.org/ (accessed 21/04/2008).

[26]   Enurgi, Home Page, available at http://www.enurgi.com/ (accessed 8/05/2008).

[27]   TauMed, Home Page, available at http://www.taumed.com/ (accessed 8/05/2008).

[28]   Revolution Health, Home Page, available at http://www.revolutionhealth.com/ (accessed 8/05/2008).

[29]   iMedix, Home Page, available at http://www.imedix.com/ (accessed 8/05/2008).

[30]   S. Wildish, Personal Home Page, available at http://www.straightfrommybrain.com/indexhibitv069/.

[31]   iGuard, Home Page, available at http://www.iguard.org / (accessed 8/05/2008).

[32]   iGuard, Safety Checks, available at http://www.iguard.org/help/patients/check.html (accessed 8/05/2008).

[33]   Nintendo WiiFit, Home Page, available at http://www.nintendo.com/wiifit/en/ (accessed 8/05/2008).

[34]   B. J. Rhodes, "The wearable remembrance agent: A system for augmented memory," *Personal Technologies Journal*, 1997

[35]   L. Srivastava, "Mobile mania, mobile manners," in *Thumb Culture: The Meaning of Mobile Phones for Society.* (P. Glotz, S. Bertschi, and C. Locke eds.), New Brunswick: Transaction Publishers, 2005.

[36]   D. Hurley, "Pole star: Human rights in the information society," September 2003, available at http://www.ichrdd.ca/english/commdoc/publications/globalization/wsis/PoleStar-Eng.html (accessed 21/07/08).

[37]   J. Austen, *Sense and Sensibility.* 1811.

[38]   N. Postman, *Technopoly: The Surrender of Culture to Technology.* New York: Vintage Books, 1993.

[39]   L. Srivastava, "The dissemination and acquisition of knowledge in the mobile age," in *Mobile Understanding: The epistemology of Ubiquitous Communication.* (K. Nyiri ed.), Vienna: Passagen Verlag, 2006.

[40]   O. Wilde, *The Soul of Man Under Socialism Socialism.* 1891.

Index

*For Product Safety Concerns and Information please contact
our EU representative GPSR@taylorandfrancis.com Taylor & Francis
Verlag GmbH, Kaufingerstraße 24, 80331 München, Germany*

T - #0127 - 230425 - C196 - 234/156/9 [11] - CB - 9788792329226 - Gloss Lamination